BRITISH RAILW

COA
STOCK

THIRTY-SECOND EDITION
2008

The complete guide to all
Locomotive-Hauled Coaches which
operate on National Rail

Peter Hall, Peter Fox & Robert Pritchard

ISBN 978 1902 336 60 2

© 2007. Platform 5 Publishing Ltd., 3 Wyvern House, Sark Road, Sheffield, S2 4HG, England.

CONTENTS

PROVISION OF INFORMATION

This book has been compiled with care to be as accurate as possible, but in some cases official information is not available and the publisher cannot be held responsible for any errors or omissions. We would like to thank the companies and individuals which have been co-operative in supplying information to us. The authors of this series of books are always pleased to receive notification from readers of any inaccuracies readers may find in the series, to enhance future editions. Please send comments to:

Robert Pritchard, Platform 5 Publishing Ltd., 3 Wyvern House, Sark Road, Sheffield, S2 4HG, England.

Tel: 0114 255 2625
Fax: 0114 255 2471
e-mail: robert@platform5.com

This book is updated to 10 October 2007.

UPDATES

This book is updated to the Stock Changes given in Today's Railways UK 72 (December 2007). Readers are therefore advised to update this book from the official Platform 5 Stock Changes published every month in **Today's Railways UK** magazine, starting with issue 73.

The Platform 5 magazine **Today's Railways UK** contains news and rolling stock information on the railways of Britain and Ireland and is published on the second Monday of every month. For further details of **Today's Railways UK**, please see the advertisement on the back cover of this book.

BRITAIN'S RAILWAY SYSTEM

INFRASTRUCTURE & OPERATION

Britain's national railway infrastructure is owned by a "not for dividend" company, Network Rail. Many stations and maintenance depots are leased to and operated by Train Operating Companies (TOCs), but some larger stations remain under Network Rail control. The only exception is the infrastructure on the Isle of Wight, which is nationally owned and is leased to South West Trains.

Trains are operated by TOCs over Network Rail, regulated by access agreements between the parties involved. In general, TOCs are responsible for the provision and maintenance of the locomotives, rolling stock and staff necessary for the direct operation of services, whilst NR is responsible for the provision and maintenance of the infrastructure and also for staff needed to regulate the operation of services.

DOMESTIC PASSENGER TRAIN OPERATORS

The large majority of passenger trains are operated by the TOCs on fixed term franchises. Franchise expiry dates are shown in parentheses in the list of franchisees below:

Franchise	Franchisee	Trading Name
Central Trains[1]	National Express Group plc (until 11 November 2007)	Central Trains
Chiltern Railways	M40 Trains Ltd. (until 1 March 2022)	Chiltern Railways
Cross-Country[2]	Virgin Rail Group Ltd. (until 11 November 2007)	Virgin Trains
Gatwick Express[3]	National Express Group plc (until 22 June 2008)	Gatwick Express
Greater Western[4]	First Group plc (until 1 April 2013)	First Great Western
Greater Anglia[5]	National Express Group plc (until 1 April 2011)	"One"
Integrated Kent[6]	GoVia Ltd. (Go-Ahead/Keolis) (until 1 April 2012)	Southeastern
InterCity East Coast[7]	GNER Holdings Ltd. (until 9 December 2007)	Great North Eastern Railway
InterCity West Coast	Virgin Rail Group Ltd. (until 9 March 2012)	Virgin Trains
LTS Rail	National Express Group plc (until 25 May 2011)	c2c
Merseyrail Electrics[8]	Serco/NedRail (until 20 July 2028)	Merseyrail Electrics

Midland Main Line[9]	National Express Group plc (until 11 November 2007)	Midland Mainline
North London Railways[10]	National Express Group plc (until 11 November 2007)	Silverlink Train Services
Northern Rail[11]	Serco/NedRail (until 12 September 2013)	Northern
ScotRail	First Group plc (until 17 October 2011)	First ScotRail
South Central[12]	GoVia Ltd. (Go-Ahead/Keolis) (until 30 September 2009)	Southern
South Western[13]	Stagecoach Holdings plc (until 4 February 2014)	South West Trains
Thameslink/Great Northern[14]	First Group plc (until 1 April 2012)	First Capital Connect
Trans-Pennine Express	First Group/Keolis (until 1 February 2012)	First Trans-Pennine Express
Wales & Borders	Arriva Trains Ltd. (until 6 December 2018)	Arriva Trains Wales

Notes:

[1] Due to be abolished on expiry. Services to be split between the new East Midlands franchise (held by Stagecoach under the brand name East Midlands Trains, also incorporating the existing Midland Mainline franchise) and the new West Midlands franchise (held by GoVia and branded London Midland, including all existing West Midlands area Central Trains and Silverlink services).

[2] The new expanded Cross-Country franchise, held by Arriva, will come into being from 11 November 2007, also including the existing Central Trains Birmingham–Stansted Airport and Nottingham–Cardiff services.

[3] Gatwick Express is to be absorbed by Southern as part of the DfT's Brighton Main Line Route Utilisation Strategy. This will now take place before the original expiry date of 2011 for the current Gatwick Express franchise.

[4] The new Greater Western franchise started on 1 April 2006 and incorporates the former Great Western, Wessex Trains and Thames Trains franchises. Awarded for seven years to 2013 with a possible extension by a further three if performance targets are met.

[5] Incorporates the former Anglia and Great Eastern franchises and the West Anglia half of West Anglia Great Northern. Awarded for seven years with a likely extension for a further three.

[6] The new Integrated Kent franchise started on 1 April 2006 for an initial period of six years to 2012, to be extended by a further two if performance targets are met.

[7] The new East Coast franchise is due to start on 9 December 2007, held by National Express Group. The current franchise started on 1 May 2005 and was intended to last for an initial period of seven years, to be extended by a further three if performance targets were met. However, the recent financial difficulties of GNER's holding company Sea Containers led to a decision by the Department for Transport to relet the franchise.

[8] Now under control of Merseyside PTE instead of the DfT. Franchise due to be reviewed after seven years and then every five years to fit in with the Merseyside Local Transport Plan.

[9] Due to be replaced by the new East Midlands franchise (East Midlands Trains) (see also Central Trains).

[10] Due to be abolished on expiry. Services to be split between the new London Rail franchise (held by MTR/Laing, under the control of Transport for London) and the new West Midlands franchise.

[11] The Northern franchise runs for up to 8¾ years.

[12] The South Central franchise termination date has been brought forward by three months to ensure that the winner of the new franchise has already taken over before any major timetable changes are made in December 2009.

[13] The new South Western franchise started on 4 February 2007, incorporating the previous South West Trains and Island Line franchises. Awarded for seven years to 2014 with a possible extension by a further three if performance targets are met.

[14] Incorporates the former Thameslink franchise and Great Northern half of the former West Anglia Great Northern franchise. Runs for six years to 2012 with a possible extension for up to three years depending on performance targets.

All new franchises officially start at 02.00 on the first day. Because of this the finishing date of an old franchise and the start date of its successor are the same.

Where termination dates are dependent on performance targets being met, the earliest possible termination date is generally given. However, in the case of Chiltern and Northern, the termination dates are based on the maximum franchise length.

The following operators run non-franchised services only:

Operator	Trading Name	Route
BAA	Heathrow Express	London Paddington–Heathrow Airport
ECT Mainline Rail/ Victa Westlink Rail	"Butlins Express"	Minehead–Bristol Temple Meads*
Hull Trains†	Hull Trains	London King's Cross–Hull
West Coast Railway Company	West Coast Railway Company	Birmingham–Stratford-upon-Avon Fort William–Mallaig* York–Scarborough*

* Special summer-dated services only.
† Owned mainly by First Group

In addition, two further open access services have been approved, but had not started operating at the time of going to press: a Sunderland–London King's Cross service operated by Grand Central (expected to start in late 2007), and a Wrexham–London Marylebone service run by the Wrexham, Shropshire & Marylebone Railway (expected to start in spring 2008).

INTERNATIONAL PASSENGER OPERATIONS

Eurostar (UK) operates passenger services between the UK and mainland Europe, jointly with the national operators of France (SNCF) and Belgium (SNCB/NMBS). Eurostar (UK) is a subsidiary of London & Continental Railways, which is jointly owned by National Express Group and British Airways.

In addition, a service for the conveyance of accompanied road vehicles through the Channel Tunnel is provided by the tunnel operating company, Eurotunnel.

FREIGHT TRAIN OPERATIONS

The following operators operate freight train services under "Open Access" arrangements:

English Welsh & Scottish Railway (EWS)
Freightliner
Direct Rail Services
First GBRf

Fastline (Jarvis)
Victa Westlink Rail
Advenza (Cotswold Rail)
Colas Rail

INTRODUCTION

NUMBERING SYSTEMS

Seven different numbering systems were in use on BR. These were the BR series, the four pre-nationalisation companies' series', the Pullman Car Company's series and the UIC (International Union of Railways) series. BR number series coaches and former Pullman Car Company series are listed separately. There is also a separate listing of 'Saloon' type vehicles which are registered to run on the national railway system. Please note that the Mark 2 Pullman vehicles were ordered after the Pullman Car Company had been nationalised and are therefore numbered in the BR series.

In previous issues of this book, non-passenger-carrying coaching stock was listed in a separate section. However, since the only non-passenger-carrying vehicles still in service are the Mark 3 and Mark 4 driving brake vans ("driving van trailers") and two full kitchen cars, all of which run in passenger trains, these are now included with the BR passenger-carrying stock.

LAYOUT OF INFORMATION

Coaches are listed in numerical order of painted number in batches according to type.

Each coach entry is laid out as in the following example (previous number(s) column may be omitted where not applicable):

No.	Prev. No.	Notes	Livery	Owner	Operator	Depot/Location
10229	(11059)	*	1	P	1	NC

Note that the operator is the organisation which facilitates the use of the coach and may not be the actual train operating company which runs the train.

DETAILED INFORMATION & CODES

Under each type heading, the following details are shown:

- "Mark" of coach (see below).
- Descriptive text.
- Number of first class seats, standard class seats, lavatory compartments and wheelchair spaces shown as F/S nT nW respectively.
- Bogie type (see below).
- Additional features.
- ETH Index.

TOPS TYPE CODES

TOPS type codes are allocated to all coaching stock. For vehicles numbered in the passenger stock number series the code consists of:

(1) Two letters denoting the layout of the vehicle as follows:

AA   Gangwayed Corridor
AB   Gangwayed Corridor Brake
AC   Gangwayed Open (2+2 seating)
AD   Gangwayed Open (2+1 seating)
AE   Gangwayed Open Brake
AF   Gangwayed Driving Open Brake
AG   Micro-Buffet
AH   Brake Micro-Buffet
AI   As "AC" but with drop-head buckeye and gangway at one end only
AJ   Kitchen or Buffet Car with seating -
AK   Kitchen Car
AL   As "AC" but with disabled person's toilet (Mark 4 only)
AN   Open Second with Miniature Buffet
AP   Pullman Kitchen with Servery
AQ   Pullman Parlour First
AR   Pullman Brake First
AS   Sleeping Car
AT   Royal Train Coach
AU   Sleeping Car with Pantry
AX   Generator Van
AZ   Special Saloon
NW   Desiro Barrier Vehicle
NZ   Driving Brake Van ("Driving Van Trailer")

(2) A digit denoting the class of passenger accommodation:

1   First
2   Standard (formerly second)
3   Composite (first & standard)
4   Unclassified
5   None

(3) A suffix relating to the build of coach.

1   Mark 1
Z   Mark 2
A   Mark 2A
B   Mark 2B
C   Mark 2C
D   Mark 2D
E   Mark 2E
F   Mark 2F
G   Mark 3 or 3A
H   Mark 3B
J   Mark 4

OPERATING CODES

Operating codes used by train company operating staff (and others) to denote vehicle types in general. These are shown in parentheses adjacent to TOPS type codes. Letters used are:

B Brake	K Side corridor with lavatory
C Composite	O Open
F First Class	S Standard Class (formerly second)

Various other letters are in use and the meaning of these can be ascertained by referring to the titles at the head of each type.

Readers should note the distinction between an SO (Open Standard) and a TSO (Tourist Open Standard) The former has 2+1 seating layout, whilst the latter has 2+2.

BOGIE TYPES

BR Mark 1 (BR1). Double bolster leaf spring bogie. Generally 90 m.p.h., but Mark 1 bogies may be permitted to run at 100 m.p.h. with special maintenance. Weight: 6.1 t.

BR Mark 2 (BR2). Single bolster leaf-spring bogie used on certain types of non-passenger stock and suburban stock (all now withdrawn). Weight: 5.3 t.

COMMONWEALTH (C). Heavy, cast steel coil spring bogie. 100 m.p.h. Weight: 6.75 t.

B4. Coil spring fabricated bogie. Generally 100 m.p.h., but B4 bogies may be permitted to run at 110 m.p.h. with special maintenance. Weight: 5.2 t.

B5. Heavy duty version of B4. 100 m.p.h. Weight: 5.3 t.
B5 (SR). A bogie originally used on Southern Region EMUs, similar in design to B5. Now also used on locomotive hauled coaches. 100 m.p.h.
BT10. A fabricated bogie designed for 125 m.p.h. Air suspension.
T4. A 125 m.p.h. bogie designed by BREL (now Bombardier Transportation).
BT41. Fitted to Mark 4 vehicles, designed by SIG in Switzerland. At present limited to 125 m.p.h., but designed for 140 m.p.h.

BRAKES

Air braking is now standard on British main line trains. Vehicles with other equipment are denoted:

v   Vacuum braked.
x   Dual braked (air and vacuum).

HEATING & VENTILATION

Electric heating and ventilation is now standard on British main-line trains. Certain coaches for use on charter services may also have steam heating facilities, or be steam heated only.

PUBLIC ADDRESS

It is assumed all coaches are now fitted with public address equipment, although certain stored vehicles may not have this feature. In addition, it is assumed all vehicles with a conductor's compartment have public address transmission facilities, as have catering vehicles.

COOKING EQUIPMENT

It is assumed that Mark 1 catering vehicles have gas powered cooking equipment, whilst Mark 2, 3 and 4 catering vehicles have electric powered cooking equipment unless stated otherwise.

ADDITIONAL FEATURE CODES

d	Secondary door locking.
dg	Driver–Guard communication equipment.
f	Facelifted or fluorescent lighting.
k	Composition brake blocks (instead of cast iron).
n	Day/night lighting.
pg	Public address transmission and driver-guard communication.
pt	Public address transmission facility.
q	Catering staff to shore telephone.
w	Wheelchair space.
z	Disabled persons' toilet.
★	Blue star multiple working cables fitted.

Standard class coaches with wheelchair space also have one tip-up seat per space.

NOTES ON ETH INDICES

The sum of ETH indices in a train must not be more than the ETH index of the locomotive. The normal voltage on British trains is 1000 V. Suffix 'X' denotes 600 amp wiring instead of 400 amp. Trains whose ETH index is higher than 66 must be formed completely of 600 amp wired stock. Class 33 and 73 locomotives cannot provide a suitable electric train supply for Mark 2D, Mark 2E, Mark 2F, Mark 3, Mark 3A, Mark 3B or Mark 4 coaches. Class 55 locomotives provide an e.t.s. directly from one of their traction generators into the train line. Consequently voltage fluctuations can result in motor-alternator flashover. Thus these locomotives are not suitable for use with Mark 2D, Mark 2E, Mark 2F, Mark 3, Mark 3A, Mark 3B or Mark 4 coaches unless modified motor-alternators are fitted. Such motor alternators were fitted to Mark 2D and 2F coaches used on the East Coast main line, but few remain fitted.

BUILD DETAILS

Lot Numbers
Vehicles ordered under the auspices of BR were allocated a lot (batch) number when ordered and these are quoted in class headings and sub-headings.

Builders
These are shown in class headings, the following designations being used:

Ashford	BR, Ashford Works.
BRCW	Birmingham Railway Carriage & Wagon Company., Smethwick, Birmingham.
BREL Derby	BREL, Derby Carriage Works (later ABB/Adtranz Derby, now Bombardier Derby).
Charles Roberts	Charles Roberts and Company., Horbury, Wakefield (later Bombardier Transportation)
Cravens	Cravens, Sheffield.
Derby	BR, Derby Carriage Works (later BREL Derby, then ABB / Adtranz Derby, now Bombardier Derby).
Doncaster	BR, Doncaster Works (later BREL Doncaster, then BRML Doncaster, then ABB/Adtranz Doncaster, now Bombardier)
Eastleigh	BR, Eastleigh Works (later BREL Eastleigh, then Wessex Traincare, and Alstom Eastleigh). Closed in 2006.
Glasgow	BR Springburn Works, Glasgow (now Alstom, Glasgow)
Gloucester	The Gloucester Railway Carriage & Wagon Co.
Hunslet-Barclay	Hunslet Barclay, Kilmarnock Works
Metro-Cammell	Metropolitan-Cammell, Saltley, Birmingham (later GEC-Alsthom Birmingham, then Alstom Birmingham).
Pressed Steel	Pressed Steel, Linwood.
Swindon	BR Swindon Works
Wolverton	BR Wolverton Works (later BREL Wolverton then Railcare, Wolverton, now Alstom Wolverton).
York	BR, York Carriage Works (later BREL York, then ABB York).

Information on sub-contracting works which built parts of vehicles e.g. the underframes etc. is not shown. In addition to the above, certain vintage Pullman cars were built or rebuilt at the following works:

Metropolitan Carriage & Wagon Company, Birmingham (Now Alstom)
Midland Carriage & Wagon Company, Birmingham
Pullman Car Company, Preston Park, Brighton
Conversions have also been carried out at the Railway Technical Centre, Derby, LNWR, Crewe and Blakes Fabrications, Edinburgh.

VEHICLE NUMBERS

Where a coach has been renumbered, the former number is shown in parentheses. If a coach has been renumbered more than once, the original number is shown first in parentheses, followed by the most recent previous number. Where the former number of a coach due to be converted or renumbered is known and the conversion and/or renumbering has not yet taken place, the coach is listed under both current number (with depot allocation) and under new number (without allocation).

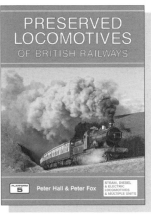

THE DEVELOPMENT OF BR STANDARD COACHES

Mark 1

The standard BR coach built from 1951 to 1963 was the Mark 1. This type features a separate underframe and body. The underframe is normally 64 ft. 6 in. long, but certain vehicles were built on shorter (57 ft.) frames. Tungsten lighting was standard and until 1961, BR Mark 1 bogies were generally provided. In 1959 Lot No. 30525 (TSO) appeared with fluorescent lighting and melamine interior panels, and from 1961 onwards Commonwealth bogies were fitted in an attempt to improve the quality of ride which became very poor when the tyre profiles on the wheels of the BR1 bogies became worn. Later batches of TSO and BSO retained the features of Lot No. 30525, but compartment vehicles – whilst utilising melamine panelling in standard class – still retained tungsten lighting. Wooden interior finish was retained in first class vehicles where the only change was to fluorescent lighting in open vehicles (except Lot No. 30648, which had tungsten lighting). In later years many Mark 1 coaches had BR 1 bogies replaced by B4.

XP64

In 1964, a new prototype train was introduced. Known as 'XP64', it featured new seat designs, pressure heating & ventilation, aluminium compartment doors and corridor partitions, foot pedal operated toilets and B4 bogies. The vehicles were built on standard Mark 1 underframes. Folding exterior doors were fitted, but these proved troublesome and were later replaced with hinged doors. All XP64 coaches have been withdrawn, but some have been preserved.

Mark 2

The prototype Mark 2 vehicle (W 13252) was produced in 1963. This was an FK of semi-integral construction and had pressure heating & ventilation, tungsten lighting, and was mounted on B4 bogies. This vehicle has been preserved by the National Railway Museum and is currently stored at MoD Kineton DM. The production build was similar, but wider windows were used. The TSO and SO vehicles used a new seat design similar to that in the XP64 and fluorescent lighting was provided. Interior finish reverted to wood. Mark 2 vehicles were built from 1964–66.

Mark 2A–2C

The Mark 2A design, built 1967–68, incorporated the remainder of the features first used in the XP64 coaches, i.e. foot pedal operated toilets (except BSO), new first class seat design, aluminium compartment doors and partitions together with fluorescent lighting in first class compartments. Folding gangway doors (lime green coloured) were used instead of the traditional one-piece variety.

Mark 2B coaches had wide wrap around doors at vehicle ends, no centre doors and a slightly longer body. In standard class there was one toilet at each end instead of two at one end as previously. The folding gangway doors were red.

Mark 2C coaches had a lowered ceiling with twin strips of fluorescent lighting and ducting for air conditioning, but air conditioning was never fitted.

Mark 2D–2F

These vehicles were fitted with air conditioning. They had no opening top-lights in saloon windows, which were shallower than previous ones.

Mark 2E vehicles had smaller toilets with luggage racks opposite. The folding gangway doors were fawn coloured.

Mark 2F vehicles had a modified air conditioning system, plastic interior panels and Inter-City 70 type seats.

Mark 3

The Mark 3 design has BT10 bogies, is 75 ft. (23 m.) long and is of fully integral construction with Inter-City 70 type seats. Gangway doors were yellow (red in RFB) when new, although these were changed on refurbishment. Loco-hauled coaches are classified Mark 3A, Mark 3 being reserved for HST trailers. A new batch of FO and BFO, classified Mark 3B, was built in 1985 with Advanced Passenger Train-style seating and revised lighting. These are now in use on First Great Western sleeping car trains. The last vehicles in the Mark 3 series were the driving brake vans ("driving van trailers") built for West Coast Main Line services.

A number of Mark 3 vehicles are now being converted for use as HST trailers with Grand Central and Arriva Cross-Country.

Mark 4

The Mark 4 design was built by Metro-Cammell for use on the East Coast Main Line after electrification and featured a body profile suitable for tilting trains, although tilt is not fitted, and is not intended to be. This design is suitable for 140 m.p.h. running, although is restricted to 125 m.p.h. because the signalling system on the route is not suitable for the higher speed. The bogies for these coaches were built by SIG in Switzerland and are designated BT41. Power operated sliding plug exterior doors are standard. All Mark 4s were rebuilt with completely new interiors in 2003–05 and were referred to as "Mallard" stock by GNER.

▲ **Passenger Carrying Coaching Stock.** Chocolate & Cream-liveried Mark 1 Buffet Standard 1863 is seen at Heeley, Sheffield on 29/08/07. **Robert Pritchard**

▼ Carmine & Cream-liveried Mark 1 FO 3119 (with Commonwealth bogies), owned by Riviera Trains, in the formation of the Railway Touring Company's "Autumn Highlander" charter at Doncaster on 05/10/07. **Peter Fox**

▲ BR Maroon-liveried 3141, with EWS branding, is seen at Bristol Temple Meads on 12/07/07. **Robert Pritchard**

▼ Newly repainted into West Coast Railway Company Maroon livery, Mark 1 TSO 5035 is seen at Aberystwyth on 18/10/06. **Robert Pritchard**

▲ Mark 1 Generator Van 6312, in Pullman livery, is seen at Aberystwyth on 18/10/06. **Robert Pritchard**

▼ BR Green-liveried Mark 1 BFK 17015 is seen at Heeley, Sheffield on 29/08/07. **Robert Pritchard**

▲ Pullman Car Company-liveried Mark 2 Pullman Parlour First 548 "GRASMERE" is seen at Aberystwyth on 18/10/06. These vehicles were built for the Manchester and Liverpool Pullmans. **Robert Pritchard**

▼ Chocolate & Cream-liveried Mark 2A TSO 5350 is seen passing Ancaster on 11/08/07. **Robert Pritchard**

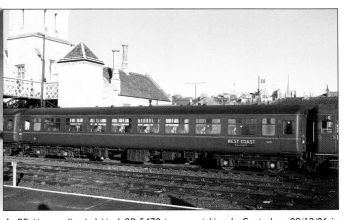

▲ BR Maroon-liveried Mark 2B 5478 is seen at Lincoln Central on 09/12/06 in the formation of an additional service in connection with the Lincoln Christmas Market. **Robert Pritchard**

▼ Riviera Trains Oxford blue-liveried Mark 2B Couchette/Generator Coach 17105 is seen near Buxton on 27/06/07. **Robert Pritchard**

▲ West Coast Railway Company Maroon-liveried Mark 2C 5569 is seen near Hathersage on 14/07/07. **Robert Pritchard**

▼ Riviera Trains-liveried Mark 2F FO 3390 "CONSTABLE", which forms part of Riviera's "Great Briton" set, is seen near Buxton on 27/06/07. **Robert Pritchard**

▲ Blue & Grey-liveried Mark 3A RFM 10246 is seen at Kensington Olympia on 24/07/07. This coach is owned and operated by Cargo-D and used on charter services. **Iain Scotchman**

▼ DRS-owned and liveried Mark 3A FO 11019 is seen at Carlisle Kingmoor depot on 07/07/07. **Robert Pritchard**

▲ "One"-liveried Mark 3A TSO 12120 is seen at Chelmsford on 04/06/07.
Robert Pritchard

▼ First Group-liveried Mark 3A "SLEP" Sleeping Car 10520 is seen at London Euston on 22/08/07.
Robert Pritchard

1. BR NUMBER SERIES COACHING STOCK

AJ11 (RF)   KITCHEN FIRST

Mark 1. Spent most of its life as a Royal Train vehicle and was numbered 2907 for a time. Built with Commonwealth bogies, but B5 bogies substituted. 24/–. ETH 2.

Lot No. 30633 Swindon 1961. 41 t.

325	**PC**	VS	*VS*	SL

AP1Z (PFK) PULLMAN KITCHEN WITH SERVERY

Mark 2. Pressure Ventilated. Seating removed and replaced with servery. 2T. B5 bogies. ETH 6.

Lot No. 30755 Derby 1966. 40 t.

504	**PC**	WC	*WC*	CS	ULLSWATER
506	**PC**	WC	*WC*	CS	WINDERMERE

AQ1Z (PFP)   PULLMAN PARLOUR FIRST

Mark 2. Pressure Ventilated. 36/– 2T. B4 bogies. ETH 5.

Non-standard livery: 546 is maroon & beige.

Lot No. 30754 Derby 1966. 35 t.

546	**0**	WC		CS	CITY OF MANCHESTER
548	**PC**	WC	*WC*	CS	GRASMERE
549	**PC**	WC	*WC*	CS	BASSENTHWAITE
550	**PC**	WC	*WC*	CS	RYDAL WATER
551	**PC**	WC	*WC*	CS	BUTTERMERE
552	**PC**	WC	*WC*	CS	ENNERDALE WATER
553	**PC**	WC	*WC*	CS	CRUMMOCK WATER

AR1Z (PFB)   PULLMAN BRAKE FIRST

Mark 2. Pressure Ventilated. 30/– 2T. B4 bogies. ETH 4.

Lot No. 30753 Derby 1966. 35 t.

586	**PC**	WC	*WC*	CS	DERWENTWATER

AJ1F (RFB) BUFFET FIRST

Mark 2F. Air conditioned. Converted 1988–9/91 at BREL, Derby from Mark 2F FOs. 1200/3/11/14–16/20/21/50/2/5/6/9 have Stones equipment, others have Temperature Ltd. 25/– 1T 1W (except 1253 which is 26/– 1T). B4 bogies. d ETH 6X.

1200/3/11/14/16/20/52/5/6. Lot No. 30845 Derby 1973. 33 t.
1201/4/5/7/8/10/12/13/15/19/21/50/1/4/8/60. Lot No. 30859 Derby 1973–74 33 t.
1202/53/9. Lot No. 30873 Derby 1974–75. 33 t.

† Fitted with new m.a. sets.

1200	(3287, 6459)		**RV**	H	RV	CP
1201	(3361, 6445)		**CH**	H	VT	TM
1202	(3436, 6456)	†	**V**	H		KT
1203	(3291)	†		H	RV	CP
1204	(3401)	†	**V**	H		SN
1205	(3329, 6438)	†	**V**	DE		ZA
1207	(3328, 6422)	†	**V**	H		KT
1208	(3393)		**V**	H		KT
1210	(3405, 6462)	†	**FS**	H	SR	IS
1211	(3305)			H	MR	BH
1212	(3427, 6453)	†	**V**	H	RV	CP
1213	(3419)	†	**V**	DM		MQ
1214	(3317, 6433)		**AR**	H		KT
1215	(3377)		**AR**	H		KT
1216	(3302)	†	**V**	H		KT
1219	(3418)		**AR**	H		KT
1220	(3315, 6432)	†	**FS**	H	SR	IS
1221	(3371)			H		KT
1250	(3372)	†	**V**	H	RV	CP
1251	(3383)	†	**V**	H		KT
1252	(3280)	†	**V**	H		LM
1253	(3432)	†	**V**	H		KT
1254	(3391)	†	**V**	H		LM
1255	(3284)	†	**V**	H		KT
1256	(3296)	†		H		MQ
1258	(3322)	†	**V**	H		KT
1259	(3439)	†	**V**	H		KT
1260	(3378)	†	**V**	H	RV	CP

AK51 (RKB) KITCHEN WITH BAR

Mark 1. Built with no seats but three Pullman-style seats now fitted in bar area. B5 bogies. ETH 1.

Lot No. 30624 Cravens 1960–61. 41 t.

1566		**VN**	VS	VS	CP

AJ41 (RBR) UNCLASSIFIED KITCHEN BUFFET

Mark 1. Built with 23 loose chairs. All remaining vehicles refurbished with 23 fixed polypropylene chairs and fluorescent lighting. ETH 2 (2X*). 1683/91/92/99 were further refurbished with 21 chairs, wheelchair space and carpets.

s Modified for use as servery vehicle with some or all seating removed.
t Modified with 11 chairs with a food preparation area replacing the former seating area.

1651–1699. Lot No. 30628 Pressed Steel 1960–61. Commonwealth bogies. 39 t.
1730. Lot No. 30512 BRCW 1960–61. B5 bogies. 37 t.

Non-standard liveries: 1683 and 1699 Oxford Blue.
1657 Nanking blue.
1679, 1680 & 1698 British racing green & cream lined out in gold.

1651	t	**CC**	R V	*RV*	CP	1683		**O**	R V	*RV*	CP
1657	s	**O**	C D	*CD*	EM	1691	t	**CC**	R V	*RV*	CP
1658	t	**BG**	E		OM	1692	t	**CH**	R V	*RV*	CP
1659	s	**PC**	R A	*WT*	OM	1696	t	**G**	E	*E*	OM
1671	x*t	**M**	R V	*RV*	CP	1698	s	**O**	E		OM
1679	t	**O**	E	*E*	OM	1699	t	**O**	R V	*RV*	CP
1680	*t	**O**	E	*E*	OM	1730	x	**M**	B K	*BK*	BT

AN2F (RSS) SELF-SERVICE BUFFET

Mark 2F. Air conditioned. Temperature Ltd. equipment. Inter-City 70 seats. Converted 1974 from a Mark 2F TSO as a prototype self-service buffet for APT-P. Sold to Northern Ireland Railways 1983 and regauged to 5'3". Since withdrawn, repatriated to Great Britain and converted back to standard gauge. –/24. B5 bogies. ETH 12X.

Lot No. 30860 Derby 1973–74. 33 t.

1800 (5970, NIR546) **PC** WT *WT* OM

AN21 (RMB)
OPEN STANDARD WITH MINIATURE BUFFET

Mark 1. –/44 2T. These vehicles are basically an open standard with two full window spaces removed to accommodate a buffet counter, and four seats removed to allow for a stock cupboard. All remaining vehicles now have fluorescent lighting. Commonwealth bogies. ETH 3.

1813–1832. Lot No. 30520 Wolverton 1960. 38 t.
1840–1842. Lot No. 30507 Wolverton 1960. 37 t.
1859–1863. Lot No. 30670 Wolverton 1961–62. 38 t.
1882. Lot No. 30702 Wolverton 1962. 38 t.

Notes:

1823 is registered for use between Middlesbrough and Whitby only.

1842 is refurbished and fitted with a microwave oven.

1861 has had its toilets replaced with store cupboards.

1813	x	**M**	R V	*RV*	CP		1859	x	**M**	B K	*BK*	BT
1823	v	**M**	N Y	*NY*	NY		1860	x	**M**	WC	*WC*	CS
1832	x	**G**	R V		OM		1861	x	**M**	WC	*WC*	CS
1840	v	**G**	WC	*WC*	CS		1863	x	**CH**	R V	*RV*	CP
1842		**CH**	R V	*RV*	CP		1882	x	**M**	WC	*WC*	CS

AJ41 (RBR)   UNCLASSIFIED KITCHEN BUFFET

Mark 1. These vehicles were built as unclassified restaurant (RU). They were rebuilt with buffet counters and 23 fixed polypropylene chairs (RBS), then further refurbished by fitting fluorescent lighting and reclassified RBR. ETH 2X.

s Modified for use as servery vehicle with seating removed.

1953. Lot No. 30575 Swindon 1960. B4/B5 bogies. 36.5 t.
1961. Lot No. 30632 Swindon 1961. Commonwealth bogies. 39 t.

1953	s	**VN**	V S	*VS*	CP		1961	x	**G**	WC	*WC*	CS

AU51   CHARTER TRAIN STAFF COACHES

Mark 1. Converted from BCKs in 1988. Commonwealth bogies. ETH 2.

Lot No. 30732 Derby 1964. 37 t.

Non-Standard livery: 2834 British racing green & cream lined out in gold.

2833	(21270)	**BG**	E		OM
2834	(21267)	**0**	R V	*RV*	CP

AT5G   HM THE QUEEN'S SALOON

Mark 3. Converted from a FO built 1972. Consists of a lounge, bedroom and bathroom for HM The Queen, and a combined bedroom and bathroom for the Queen's dresser. One entrance vestibule has double doors. Air conditioned. BT10 bogies. ETH 9X.

Lot No. 30886 Wolverton 1977. 36 t.

2903	(11001)	**RP**	N R	*RP*	ZN

AT5G HRH THE DUKE OF EDINBURGH'S SALOON

Mark 3. Converted from a TSO built 1972. Consists of a combined lounge/dining room, a bedroom and a shower room for the Duke, a kitchen and a valet's bedroom and bathroom. Air conditioned. BT10 bogies. ETH 15X.

Lot No. 30887 Wolverton 1977. 36 t.

2904	(12001)	**RP**	N R	*RP*	ZN

AT5G ROYAL HOUSEHOLD SLEEPING CAR

Mark 3A. Built to similar specification as SLE 10647–729. 12 sleeping compartments for use of Royal Household with a fixed lower berth and a hinged upper berth. 2T plus shower room. Air conditioned. BT10 bogies. ETH 11X.

Lot No. 31002 Derby/Wolverton 1985. 44 t.

2915   **RP**   N R   *RP*   ZN

AT5G HRH THE PRINCE OF WALES'S DINING CAR

Mark 3. Converted from HST TRUK built 1976. Large kitchen retained, but dining area modified for Royal use seating up to 14 at central table(s). Air conditioned. BT10 bogies. ETH 13X.

Lot No. 31059 Wolverton 1988. 43 t.

2916   (40512)   **RP**   N R   *RP*   ZN

AT5G ROYAL KITCHEN/HOUSEHOLD DINING CAR

Mark 3. Converted from HST TRUK built 1977. Large kitchen retained and dining area slightly modified with seating for 22 Royal Household members. Air conditioned. BT10 bogies. ETH 13X.

Lot No. 31084 Wolverton 1990. 43 t.

2917   (40514)   **RP**   N R   *RP*   ZN

AT5G ROYAL HOUSEHOLD CARS

Mark 3. Converted from HST TRUKs built 1976/7. Air conditioned. BT10 bogies. ETH 10X.

Lot Nos. 31083 (31085*) Wolverton 1989. 41.05 t.

2918   (40515)   **RP**   N R   ZN
2919   (40518)   *   **RP**   N R   ZN

AT5B ROYAL HOUSEHOLD COUCHETTES

Mark 2B. Converted from BFK built 1969. Consists of luggage accommodation, guard's compartment, workshop area, 350 kW diesel generator and staff sleeping accommodation. B5 bogies. ETH2X.

Lot No. 31044 Wolverton 1986. 48 t.

2920   (14109, 17109)   **RP**   N R *RP*   ZN

Mark 2B. Converted from BFK built 1969. Consists of luggage accommodation, kitchen, brake control equipment and staff accommodation. B5 bogies. ETH7X.

Lot No. 31086 Wolverton 1990. 41.5 t.

2921   (14107, 17107)   **RP**   N R *RP*   ZN

AT5H HRH THE PRINCE OF WALES'S SLEEPING CAR

Mark 3B. BT10 bogies. Air conditioned. ETH 7X.

Lot No. 31035 Derby/Wolverton 1987.

| 2922 | | **RP** | N R | *RP* | ZN |

AT5H   ROYAL SALOON

Mark 3B. BT10 bogies. Air conditioned. ETH 6X.

Lot No. 31036 Derby/Wolverton 1987.

| 2923 | | **RP** | N R | *RP* | ZN |

AD11 (FO)   OPEN FIRST

Mark 1. 42/– 2T. ETH 3. Many now fitted with table lamps.

Non-Standard livery: 0 British racing green & cream lined out in gold.

3066–3069. Lot No. 30169 Doncaster 1955. B4 bogies. 33 t.
3096–3100. Lot No. 30576 BRCW 1959. B4 bogies. 33 t.

3068 was numbered DB 975606 for a time when in departmental service for BR.

3066		**CC**	R V	*RV*	CP		3097		**0**	R V	*RV*	CP
3068		**CC**	R V	*RV*	CP		3098	x **CH**		R V	*RV*	CP
3069		**CC**	R V	*RV*	CP		3100	x **M**		E		OM
3096	x **M**		B K	*BK*	BT							

Later design with fluorescent lighting, aluminium window frames and Commonwealth bogies.

3105–3128. Lot No. 30697 Swindon 1962–63. 36 t.
3130–3150. Lot No. 30717 Swindon 1963. 36 t.

3128/36/41/3/4/6/7/8 were renumbered 1058/60/3/5/6/8/9/70 when reclassified RUO, then 3600/5/8/9/2/6/4/10 when declassified to SO, but have since regained their original numbers. 3136 was numbered DB977970 for a time when in use with Serco Railtest as a Brake Force Runner.

3105	x **M**	WC	*WC*	CS		3124		**G**	R V	*RV*	CP
3107	x **CH**	R V	*RV*	CP		3125	x **RV**		R V	*RV*	CP
3110	x **M**	R V	*RV*	CP		3127		**G**	E		OM
3112	x **CH**	R V	*RV*	CP		3128	x **M**		WC	*WC*	CS
3113	x **M**	WC	*WC*	CS		3130	v **M**		WC	*WC*	CS
3114		**G**	E		OM		3131	x **M**		E		OM
3115	x **M**	B K	*BK*	BT		3132	x **M**		E		OM
3117	x **M**	WC	*WC*	CS		3133	x **M**		E		OM
3119		**CC**	R V	*RV*	CP		3136		**M**	WC	*WC*	CS
3120		**0**	R V	*RV*	CP		3140	x **CH**		R V	*RV*	CP
3121		**0**	R V	*RV*	CP		3141		**M**	R V	*RV*	CP
3122	x **CH**	R V	*RV*	CP		3143		**M**	WC	*WC*	CS
3123		**0**	R V	*RV*	CP		3144	x **M**		R V	*RV*	CP

3146	**M**	R V	*RV*	CP		3149	**CC**	R V	*RV*	CP
3147	**O**	R V	*RV*	CP		3150	**G**	B K		BT
3148	**BG**	R V	*RV*	CP						

AD1D (FO)   OPEN FIRST

Mark 2D. Air conditioned. Stones equipment. 42/– 2T. B4 bogies. ETH 5.

† Interior modified to Pullman Car standards with new seating, new panelling, tungsten lighting and table lights for VSOE "Northern Belle".

Lot No. 30821 Derby 1971–72. 34 t.

| 3174 | † | **VN** | V S | *VS* | CP | | 3188 | | **PC** | R A | *WT* | OM |
| 3182 | † | **VN** | V S | *VS* | CP | | | | | | | |

AD1E (FO)   OPEN FIRST

Mark 2E. Air conditioned. Stones equipment. 42/– 2T (41/– 2T 1W w, 36/– 2T p). B4 bogies. ETH 5.

r Refurbished with new seats.
u Fitted with power supply for Mk. 1 RBR.
† Interior modified to Pullman Car standards with new seating, new panelling, tungsten lighting and table lights for VSOE "Northern Belle".

3255 was numbered 3525 for a time when fitted with a pantry.

Lot No. 30843 Derby 1972–73. 32.5 t. (35.8 t. †).

3223		**RV**	R V	*RV*	CP		3247	†	**VN**	V S	*VS*	CP
3228	du	**RV**	H	*RV*	CP		3255	dr	**M**	E		OM
3229	d	**RV**	H		KT		3261	dw	**FP**	H	*RV*	CP
3231	p	**PC**	R A	*WT*	OM		3267	†	**VN**	V S	*VS*	CP
3232	dr	**G**	V S	*VS*	CP		3269	dr	**M**	E		OM
3240		**RV**	R V	*RV*	CP		3273	†	**VN**	V S	*VS*	CP
3241	dr	**FP**	H	*CD*	GL		3275	†	**VN**	V S	*VS*	CP
3244	d	**RV**	H	*RV*	CP							

AD1F (FO)   OPEN FIRST

Mark 2F. Air conditioned. 3277–3318/58–79 have Stones equipment, others have Temperature Ltd. 42/– 2T. All now refurbished with power-operated vestibule doors, new panels and new seat trim. B4 bogies. d. ETH 5X.

3277–3318. Lot No. 30845 Derby 1973. 33.5 t.
3325–3426. Lot No. 30859 Derby 1973–74. 33.5 t.
3431–3438. Lot No. 30873 Derby 1974–75. 33.5 t.

r Further refurbished with table lamps, modified seats with burgundy seat trim and new m.a. sets.
s Further refurbished with table lamps and modified seats with burgundy seat trim.
u Fitted with power supply for Mk. 1 RBR.

| 3277 | | **AR** | H | *RV* | CP | | 3278 | r | **BP** | FM | *VI* | EM |

3279	u	**M**	E	*E*	OM	3359	s	**V**	WC	CS
3285	s	**V**	H		LM	3360	s		WC *WC*		CS
3292		**M**	E	*E*	OM	3362	s		WC *WC*		CS
3295		**AR**	H	*RV*	CP	3364	r	**RV**	RV *RV*		CP
3299	r	**V**	H		KT	3366	s	**V**	H		LM
3303		**AR**	H		OM	3368		**M**	E		OM
3304	r	**V**	H	*RV*	CP	3374			H	*MR*	BH
3309		**CH**	H	*VT*	TM	3375		**M**	E	*E*	OM
3312			H	*MR*	BH	3379	u	**AR**	H	*RV*	CP
3313	r	**BP**	CD	*CD*	EM	3384	r	**RV**	RV *RV*		CP
3314	r	**V**	H	*RV*	CP	3385	r	**BP**	FM	*VI*	EM
3318		**M**	E	*E*	OM	3386	r	**V**	RV	*RV*	CP
3325	r	**V**	H	*RV*	CP	3387	s	**V**	DM		MQ
3326	r	**BP**	CD	*CD*	EM	3388		**M**	E	*E*	OM
3330	r	**RV**	RV	*RV*	CP	3390	r	**RV**	RV *RV*		CP
3331		**M**	E	*E*	OM	3392	r	**BP**	CD	*CD*	EM
3333	r	**V**	H	*RV*	CP	3395	r	**BP**	CD	*CD*	EM
3334		**AR**	H	*RV*	CP	3397	r	**RV**	RV *RV*		CP
3336	u	**AR**	H	*RV*	CP	3399	u	**M**	E		OM
3338	u	**M**	E		OM	3400		**M**	E	*E*	OM
3340	r	**V**	H	*RV*	CP	3402	s	**V**	DM		MQ
3344	r	**V**	H	*RV*	CP	3408	s	**V**	WC		CS
3345	r	**V**	H	*RV*	CP	3411	s	**V**	DM		MQ
3348	r	**RV**	RV	*RV*	CP	3414		**M**	E	*E*	OM
3350	r	**BP**	CD	*CD*	EM	3416			H	*VT*	TM
3351		**AR**	H	*VT*	TM	3417		**AR**	H	*RV*	CP
3352	r	**BP**	CD	*CD*	EM	3424		**M**	E	*E*	OM
3353	s	**V**	DM		MQ	3425	s	**V**	DM		MQ
3354	s	**V**	DM		MQ	3426	r	**RV**	RV *RV*		CP
3356	r	**RV**	RV	*RV*	CP	3431	r	**BP**	CD	*CD*	EM
3358		**M**	E	*E*	OM	3438	s	**V**	H		LM

AC21 (TSO)   OPEN STANDARD

Mark 1. –/64 2T. ETH 4.

3766. Lot No. 30079 York 1953. Commonwealth bogies (originally built with BR Mark 1 bogies). 36 t.
3860/**3872.** Lot No. 30080 York 1954. BR Mark 1 bogies. 33 t.
4252. Lot No. 30172 York 1956. BR Mark 1 bogies. 33 t.
4290. Lot No. 30207 BRCW 1956. BR Mark 1 bogies. 33 t.
4455. Lot No. 30226 BRCW 1957. BR Mark 1 bogies. 33 t.

Notes:

3766–3872 have narrower seats than later vehicles.
3860–4455 are registered for use between Middlesbrough and Whitby only.

3766	x	**M**	WC *WC*	CS		4252	v	**BG**	NY *NY*	NY
3860	v	**M**	NY *NY*	NY		4290	v	**M**	NY *NY*	NY
3872	v	**BG**	NY *NY*	NY		4455	v	**CC**	NY *NY*	NY

AD21 (SO)   OPEN STANDARD

Mark 1. –/48 2T. ETH 4.

4786. Lot No. 30376 York 1957. 33 t.
4817. Lot No. 30473 BRCW 1959. 33 t.

Note: These vehicles are registered for use between Middlesbrough and Whitby only.

| 4786 | v | **M** | N Y | *NY* | NY | | 4817 | v | **M** | N Y | *NY* | NY |

AC21 (TSO)   OPEN STANDARD

Mark 1. These vehicles are a development of the above with fluorescent lighting and modified design of seat headrest. Built with BR Mark 1 bogies. –/64 2T. ETH 4.

4831–4836. Lot No. 30506 Wolverton 1959. Commonwealth bogies. 33 t.
4856. Lot No. 30525 Wolverton 1959–60. B4 bogies. 33 t.

| 4831 | x | **M** | B K | *BK* | BT | | 4836 | x | **M** | B K | *BK* | BT |
| 4832 | x | **M** | B K | *BK* | BT | | 4856 | x | **M** | B K | *BK* | BT |

AC21 (TSO) OPEN STANDARD

Mark 1. Later vehicles built with Commonwealth bogies. –/64 2T. ETH4.

4902–4912. Lot No. 30646 Wolverton 1961. BR Mark 1 bogies substituted by the SR. All now re-rebogied. 34 t B4, 36 t C.
4925–5044. Lot No. 30690 Wolverton 1961–62. Aluminium window frames. 37 t.

Note: 5000 and 5029 are registered for use between Middlesbrough and Whitby only.

4902	x B4	**CH**	R V	*RV*	CP		4994	x	**M**	WC	*WC*	CS
4905	x C	**M**	WC	*WC*	CS		4996	x	**M**	E		OM
4912	x C	**M**	WC	*WC*	CS		4998		**M**	R V	*RV*	CP
4925		**G**	E		OM		4999		**BG**	R V		OM
4927	x	**CH**	R V	*RV*	CP		5000		**M**	N Y	*NY*	NY
4931	v	**M**	WC	*WC*	CS		5005		**BG**	E		OM
4940	x	**M**	WC	*WC*	CS		5007		**G**	E		OM
4946	x	**M**	E		OM		5008	x	**M**	R V	*RV*	CP
4949	x	**M**	R V	*RV*	CP		5009	x	**CH**	R V	*RV*	CP
4951	x	**M**	WC	*WC*	CS		5023		**G**	R V	*RV*	CP
4954	v	**M**	WC	*WC*	CS		5027		**G**	E		OM
4956		**BG**	E		OM		5028	x	**M**	B K	*BK*	BT
4958	v	**M**	WC	*WC*	CS		5029		**CC**	N Y	*NY*	NY
4959		**BG**	E		OM		5032	x	**M**	WC	*WC*	CS
4960	x	**M**	WC	*WC*	CS		5033	x	**M**	WC	*WC*	CS
4973	x	**M**	WC	*WC*	CS		5035	x	**M**	WC	*WC*	CS
4984	x	**M**	WC	*WC*	CS		5037		**G**	E		OM
4986		**G**	R V	*RV*	CP		5040	x	**CH**	R V	*RV*	CP
4991		**BG**	E		OM		5044	x	**M**	WC	*WC*	CS

AC2Z (TSO)   OPEN STANDARD

Mark 2. Pressure ventilated. –/64 2T. B4 bogies. ETH 4.

Lot No. 30751 Derby 1965–67. 32 t.

5125	v	**G**	WC		BH	5193	v	**LN**	H		TM
5148	v	**RR**	H		TM	5194	v	**RR**	H		TM
5157	v	**CH**	H	*VT*	TM	5198	v	**CH**	E	*VT*	TM
5171	v	**G**	WC	*WC*	CS	5200	v	**G**	WC	*WC*	CS
5177	v	**CH**	H	*VT*	TM	5212	v	**LN**	H		TM
5179	v	**RR**	H		TM	5216	v	**G**	WC	*WC*	CS
5183	v	**RR**	H		TM	5221	v	**RR**	H		TM
5186	v	**RR**	H		TM	5222	v	**G**	WC	*WC*	CS
5191	v	**CH**	H	*VT*	TM						

AD2Z (SO)   OPEN STANDARD

Mark 2. Pressure ventilated. –/48 2T. B4 bogies. ETH 4.

Lot No. 30752 Derby 1966. 32 t.

5229		**PC**	WT	*WT*	OM	5239		**PC**	WT	*WT*	OM
5236	v	**G**	WC	*WC*	CS	5249	v	**G**	WC	*WC*	CS
5237	v	**G**	WC	*WC*	CS						

AC2A (TSO)   OPEN STANDARD

Mark 2A. Pressure ventilated. –/64 2T (–/62 2T w). B4 bogies. ETH 4.

5276–5341. Lot No. 30776 Derby 1967–68. 32 t.
5350–5419. Lot No. 30787 Derby 1968. 32 t.

f Facelifted vehicles.

5276	f	**RV**	RV	*RV*	CP	5350		**CH**	RV	*RV*	CP
5278		**PC**	WT	*WT*	OM	5365		**RV**	RV	*RV*	CP
5292	f	**RV**	RV	*RV*	CP	5366	f	**RV**	RV	*RV*	CP
5299		**M**	WC		CS	5376		**RV**	RV	*RV*	CP
5309		**CH**	E		OM	5386	w	**M**	E		OM
5322	f	**RV**	RV	*RV*	CP	5412	w	**M**	BK	*BK*	BT
5331		**M**	E		OM	5419	w	**PC**	WT	*WT*	OM
5341	f	**CC**	RV	*RV*	CP						

AC2B (TSO)   OPEN STANDARD

Mark 2B. Pressure ventilated. –/62 2T. B4 bogies. ETH 4.

Note: 5482 was numbered DB977936 for a time when in departmental service for British Railways.

Lot No. 30791 Derby 1969. 32 t.

5453	d	**M**	WC	*WC*	CS	5463	d	**M**	WC	*WC*	CS

| 5478 | d | **M** | WC | *WC* | CS | | 5487 | d | **M** | WC | *WC* | CS |
| 5482 | | **M** | RP | *E* | OM | | 5491 | d | **M** | WC | *WC* | CS |

AC2C (TSO)   OPEN STANDARD

Mark 2C. Pressure ventilated. –/62 2T. B4 bogies. d. ETH 4.

Lot No. 30795 Derby 1969–70. 32 t.

| 5569 | **M** | WC | *WC* | CS | |

AC2D (TSO)   OPEN STANDARD

Mark 2D. Air conditioned. Stones equipment. Refurbished with new seats and end luggage stacks. –/58 2T. B4 bogies. d. ETH 5.

Note: 5679, 5737 and 5740 are currently in use as barrier vehicles.

Lot No. 30822 Derby 1971. 33 t.

5631	**M**	E		OM		5700	**FP**	H	*CD*	GL
5632	**M**	E		OM		5710	**FP**	H	*CD*	GL
5657	**M**	E		OM		5737	**FP**	H	*GW*	OO
5669	**BP**	H	*CD*	EM		5740	**FP**	H	*GW*	OO
5679	**FP**	H	*GW*	OO						

AC2E (TSO)   OPEN STANDARD

Mark 2E. Air conditioned. Stones equipment. –/64 2T (w –/62 2T 1W). B4 bogies. d. ETH 5.

5745–5797. Lot No. 30837 Derby 1972. 33.5 t.
5810–5906. Lot No. 30844 Derby 1972–73. 33.5 t.

r Refurbished with new interior panelling.
s Refurbished with new interior panelling, modified design of seat headrest and centre luggage stack. –/60 2T (w –/58 2T 1W).
t Refurbished with new interior panelling and new seats.

5745	s	**V**	H		KT		5816	r pt		H		CD
5748	r pt		H	*RV*	CP		5821	r pt	**V**	H		KT
5750	s	**V**	H		KT		5843	rw		H		CP
5754	ws	**V**	H		KT		5853	t	**AV**	AW		CF
5769	r		H	*RV*	CP		5866	r pt★		H	*MR*	BH
5779	r		H		KT		5869	t	**AV**	AW		CF
5787	s	**V**	H		KT		5874	t	**M**	WC		CS
5788	r		H		KT		5876	s pt	**V**	H		KT
5789	r pt		H		OY		5881	ws	**V**	H		KT
5792	r		H	*RV*	CP		5886	s	**V**	H		KT
5793	wspt	**V**	H		KT		5888	wr		H		KT
5797	r★		H	*MR*	BH		5897	r		H		KT
5810	s	**V**	H		KT		5899	s	**V**	H		KT
5812	wr		H		KT		5900	wspt	**V**	H		KT
5814	r		H		CP		5901	s	**V**	H		KT
5815	ws	**V**	H		KT		5903	s	**V**	H		KT

5905 s	**V**	H		KT		5906 wspt★		H	*MR*	BH

AC2F (TSO)   OPEN STANDARD

Mark 2F. Air conditioned. Temperature Ltd. equipment. Inter-City 70 seats. All were refurbished in the 1980s with power-operated vestibule doors, new panels and new seat trim. –/64 2T. (w –/62 2T 1W) B4 bogies. d. ETH 5X.

5908–5958. Lot No. 30846 Derby 1973. 33 t.
5959–6170. Lot No. 30860 Derby 1973–74. 33 t.
6171–6184. Lot No. 30874 Derby 1974–75. 33 t.

* Early Mark 2 style seats.

These vehicles have undergone a second refurbishment with carpets and new seat trim.
r Standard refurbished vehicles with new m.a. sets.

Former Cross-Country vehicles:

s Also fitted with centre luggage stack. –/60 2T.
t Also fitted with centre luggage stack and wheelchair space. –/58 2T 1W.

Former West Coast vehicles:

u As "r" but with two wheelchair spaces. –/60 2T 2W.
† Standard refurbished vehicles with new seat trim.

No.					Depot
5908 r	**V**	H			KT
5910 u	**V**	H	*RV*		CP
5911 s	**V**	RV	*RV*		CP
5912 s	**V**	H			KT
5913 s	**AV**	AW			CF
5914 u	**V**	H			KT
5916 t		WC			CS
5917 s	**V**	WC			CS
5919 s pt	**V**	H			KT
5920 †	**V**	DM			MQ
5921	**AR**	H	*RV*		CP
5922	**M**	E			OM
5924	**M**	E			OM
5925 s pt★		H			MQ
5926		H			KT
5928	**CH**	H	*VT*		TM
5929	**AR**	H	*RV*		CP
5930 t	**V**	H			KT
5931 †w	**V**	RV	*RV*		CP
5932 r	**V**	H	*RV*		CP
5933 r	**V**	H			KT
5934 r	**V**	RV	*RV*		CP
5935	**AR**	H			KT
5936	**AR**	H			LM
5937 r	**V**	RV	*RV*		CP
5940 u	**V**	H			KT
5941 r	**V**	H			KT

No.					Depot
5943 rw	**V**	H			KT
5944	**AR**	H			KT
5945 r	**V**	H	*RV*		CP
5946 r	**V**	H	*RV*		CP
5947 s pt	**V**	H			KT
5948 u	**V**	H			KT
5949 u	**V**	H			KT
5950	**AR**	H	*RV*		CP
5952 r	**V**	RV	*RV*		CP
5954	**M**	E			OM
5955 r	**V**	RV	*RV*		CP
5957 r	**V**	H			KT
5958 s★		H		*MR*	BH
5959 n	**M**	E			OM
5960 s	**V**	H			KT
5961 s pt	**V**	RV	*RV*		CP
5962 s pt	**V**	H			KT
5963 r	**V**	H	*RV*		CP
5964	**AR**	H	*RV*		CP
5965 t	**AV**	AW			CF
5966	**AR**	H			KT
5967 t	**V**	H			ZG
5968	**AR**	H			KT
5969 u	**V**	H			KT
5971 s	**AV**	AW	*AW*		CP
5975 s	**V**	H			ZG
5976 t	**AV**	AW			CF

5977	r	**V**	H			KT
5978	r	**V**	H			KT
5980	r	**V**	H			KT
5981	s★		H			MQ
5983	s	**V**	H			KT
5984	r	**V**	H	*RV*		CP
5985		**AR**		*RV*		CP
5986	r	**V**	H	*RV*		CP
5987	r	**V**	H	*RV*		CP
5988	r	**V**	H			KT
5989	t	**V**	H			OY
5991	s	**V**	H			KT
5993	*	**AR**	H			KT
5994	r	**V**	H			KT
5995	r	**V**	H			KT
5996	s pt	**V**	H			ZG
5997	r	**V**	H	*RV*		CP
5998		**AR**	H	*RV*		CP
5999	s	**V**	H			ZG
6000	t	**V**	WC	*WC*		CS
6001	u	**V**	H			OY
6002	†	**V**	DM			MQ
6005	r	**V**	H			ZG
6006		**AR**	H	*RV*		CP
6008	s	**AV**	AW	*AW*		CP
6009	r	**V**	H			KT
6010	s	**V**	H			ZG
6011	s	**V**	H			ZG
6012	r	**V**	H			KT
6013	s	**AV**	AW			CF
6014	s pt		WC			CS
6015	t	**V**	H			ZG
6016	r	**V**	H			KT
6018	t	**V**	H			ZG
6021	r	**V**	H			KT
6022	s	**V**	WC	*WC*		CS
6024	s	**V**	RV			CD
6025	t	**V**	H			ZG
6026	s	**V**	H			ZG
6027	u	**V**	RV	*RV*		CP
6028		**AR**	H			KT
6029	r	**V**	H			KT
6031	r	**V**	H			KT
6034		**AR**	H			KT
6035	t★	**AV**	AW			CF
6036	*	**M**	E			OM
6037		**AR**	H			KT
6038	s	**V**	RV			CD
6041	s	**V**	WC	*WC*		CS
6042		**AR**	H	*RV*		CP
6043	†	**V**	RV	*RV*		CP
6045	†w	**V**	H			KT
6046	s	**V**	H			LM
6049	r	**V**	H			KT
6050	s		H			KT
6051	r	**V**	RV	*RV*		CP
6052	tw		H			KT
6053	*	**AR**	H			KT
6054	r	**V**	H	*RV*		CP
6056	†	**V**	RV	*RV*		CP
6059	s	**V**	H			KT
6061	s pt	**V**	H			ZG
6064	s	**AV**	AW	*AW*		CP
6065	r	**V**	H			KT
6066	s★	**AV**	AW			CF
6067	s pt	**V**	RV	*RV*		CP
6073	s	**V**	H			KT
6101	r	**V**	H			KT
6103		**AR**	WC	*WC*		CS
6104	r	**V**	H	*RV*		CP
6107	r	**V**	H	*RV*		CP
6110		**M**	E			OM
6111	†	**V**	RV	*RV*		CP
6112	s pt	**V**	H			ZG
6113	†	**V**	RV	*RV*		CP
6115	s		WC			CS
6117	t★	**WX**	H	*MR*		BH
6119	s	**AV**	AW			CF
6120	s	**V**	H			KT
6121	†	**V**	H			KT
6122	s★	**WX**	H	*MR*		BH
6123		**AR**	H			KT
6124	s pt★	**AV**	AW			LM
6134	†	**V**	H			KT
6135	s		WC			CS
6136	r	**V**	H			KT
6137	s pt	**AV**	AW			CF
6138	†	**V**	RV	*RV*		CP
6139	n*	**M**	E			OM
6141	u	**V**	RV	*RV*		CP
6142	†*	**V**	RV	*RV*		CP
6146	*	**AR**	H			LM
6148	s		WC			CS
6150	s		H			KT
6151	†*	**V**	H			KT
6152	*	**M**	E			OM
6153	†	**V**	H			KT
6154	r pt		H			KT
6158	r	**V**	H	*RV*		CP
6159	s pt	**V**	H			ZG
6160	*	**AR**	H			LM
6162	s pt	**AV**	AW			CF

6163	r	**V**	RV	*RV*	CP	6175	r	**V**	H			KT
6164	†	**V**	H		KT	6176	t	**V**	RV	*RV*		CP
6165	r	**V**	H		KT	6177	s	**V**	RV	*RV*		CP
6166			H		KT	6179	r	**V**	H			KT
6168	s★		H	*MR*	BH	6180	†w	**V**	RV	*RV*		CP
6170	s★	**AV**	AW		CF	6181	†wn	**V**	DM			MQ
6171	†	**V**	RV	*RV*	CP	6182	s	**V**	H			ZG
6172	s	**V**	H		ZG	6183	s	**AV**	AW			CF
6173	s★	**WX**	H	*MR*	BH	6184	s	**V**	H			ZG
6174		**AR**	H		KT							

AX51   BRAKE GENERATOR VAN

Mark 1. Renumbered 1989 from BR departmental series. Converted from NDA in 1973 to three-phase supply brake generator van for use with HST trailers. Modified 1999 for use with loco-hauled stock. B5 bogies.

Lot No. 30400 Pressed Steel 1958.

6310	(81448, 975325)	**CH**	RV	*RV*	CP

AX51   GENERATOR VAN

Mark 1. Converted from NDA in 1992 to generator vans for use on Anglo-Scottish sleeping car services. Now normally used on trains hauled by steam locomotives. B4 bogies. ETH75.

6311. Lot No. 30162 Pressed Steel 1958. 37.25 t.
6312. Lot No. 30224 Cravens 1956. 37.25 t.
6313. Lot No. 30484 Pressed Steel 1958. 37.25 t.

6311	(80903, 92911)	**B**	E		OM
6312	(81023, 92925)	**PC**	WC	*WC*	CS
6313	(81553, 92167)	**PC**	P	*VS*	SL

AG2C (TSOT)   OPEN STANDARD (TROLLEY)

Mark 2C. Converted from TSO by removal of one seating bay and replacing this by a counter with a space for a trolley. Adjacent toilet removed and converted to steward's washing area/store. Pressure ventilated. –/55 1T. B4 bogies. ETH 4.

Lot No. 30795 Derby 1969–70. 32.5 t.

6528	(5592)	**M**	WC	*WC*	CS

AN1F (RLO)   SLEEPER RECEPTION CAR

Mark 2F. Converted from FO, these vehicles consist of pantry, microwave cooking facilities, seating area for passengers (with "moveable" seats), staff toilet plus two bars. Now refurbished with new "sofa" seating as well as the "moveable" seats. Converted at RTC, Derby (6700), Ilford (6701–5) and Derby (6706–8). Air conditioned. 6700/1/3/5–8 have Stones equipment and 6702/4 have Temperature Ltd. equipment. 28/– 1T (* 30/– 1T). B4 bogies. d. ETH 5X.

6700–2/4/8. Lot No. 30859 Derby 1973–74. 33.5 t.
6703/5–7. Lot No. 30845 Derby 1973. 33.5 t.

6700	(3347)	*	**FB**	H	*SR*	IS
6701	(3346)		**FS**	H	*SR*	IS
6702	(3421)		**FS**	H	*SR*	IS
6703	(3308)		**FS**	H	*SR*	IS
6704	(3341)		**FS**	H	*SR*	IS
6705	(3310, 6430)		**FS**	H	*SR*	IS
6706	(3283, 6421)		**FS**	H	*SR*	IS
6707	(3276, 6418)		**FS**	H	*SR*	IS
6708	(3370)		**FS**	H	*SR*	IS

AN1D (RFB) BUFFET FIRST

Mark 2D. Converted from TSOT by the removal of another seating bay and fitting a proper buffet counter with boiler and microwave oven. Now converted to first class with new seating and end luggage stacks. Air conditioned. Stones equipment. 30/– 1T. B4 bogies. d. ETH 5.

Note: 6721 is currently in use as a barrier vehicle.

Lot No. 30822 Derby 1971. 33 t.

6720	(5622, 6652)	**M**	E	*E*	OM
6721	(5627, 6660)	**FP**	H	*GW*	OO
6722	(5736, 6661)	**FP**	H	*CD*	GL
6723	(5641, 6662)	**FP**	H		GL
6724	(5721, 6665)	**FP**	H		GL

AH2Z (BSOT) OPEN BRAKE STANDARD (TROLLEY)

Mark 2. These vehicles use the same body shell as the Mark 2 BFK. Converted from BSO by removal of one seating bay and replacing this with a counter with a space for a trolley. Adjacent toilet removed and converted to a steward's washing area/store. –/23 0T. B4 bogies. ETH 4.

Lot No. 30757 Derby 1966. 31 t.

9101	(9398)	v	**CH**	H	*VT*	TM
9104	(9401)	v	**G**	WC		ZA

AE21 (BSO) OPEN BRAKE STANDARD

Mark 1. –/39 1T. BR Mark 1 bogies. ETH 4.

Lot No. 30170 Doncaster 1956. 34 t.

Note: These vehicles are registered for use between Middlesbrough and Whitby only.

9267	**BG**	NY	*NY*	NY		9274	**M**	NY	*NY*	NY

AE2Z (BSO)   OPEN BRAKE STANDARD

Mark 2. These vehicles use the same body shell as the Mark 2 BFK and have first class seat spacing and wider tables. Pressure ventilated. –/31 1T. B4 bogies. ETH 4.

Lot No. 30757 Derby 1966. 31.5 t.

| 9391 | | **PC** WT *WT* | OM | | 9392 v | **G** WC *WC* | CS |

AE2A (BSO)   OPEN BRAKE STANDARD

Mark 2A. These vehicles use the same body shell as the Mark 2A BFK and have first class seat spacing and wider tables. Pressure ventilated. –/31 1T. B4 bogies. ETH 4. Modified for use as Escort Coaches.

9419. Lot No. 30777 Derby 1970. 31.5 t.
9428. Lot No. 30820 Derby 1970. 31.5 t.

| 9419 | **DR** DR *DR* | KM | | 9428 | **DR** DR *DR* | KM |

AE2C (BSO)   OPEN BRAKE STANDARD

Mark 2C. Pressure ventilated. –/31 1T. B4 bogies. ETH 4.

Lot No. 30798 Derby 1970. 32 t.

| 9440 d | **M** WC *WC* | CS | | 9448 d | **M** WC *WC* | CS |

AE2D (BSO)   OPEN BRAKE STANDARD

Mark 2D. Air conditioned (Stones). –/31 1T. B4 bogies. d. pg. ETH 5.

r Refurbished with new interior panelling.
s Refurbished with new seating –/22 1TD.
w Facelifted –/28 1W 1T.

Note: 9481 and 9490 are currently in use as barrier vehicles.

Lot No. 30824 Derby 1971. 33 t.

9479 r			H		OY		9490 s	**FP**	H	*GW*	OO
9480 w	**FP**	H			KT		9492 w	**FP**	DM		MQ
9481 s	**FP**	H	*GW*		OO		9493 s	**BP**	H	*CD*	EM
9488 s	**FP**	H			GL		9494 s	**M**	E	*E*	OM
9489 r	**V**	H			KT						

AE2E (BSO)   OPEN BRAKE STANDARD

Mark 2E. Air conditioned (Stones). –/32 1T. B4 bogies. d. pg. ETH 5.

Lot No. 30838 Derby 1972. 33 t.

r Refurbished with new interior panelling.
s Refurbished with modified design of seat headrest and new interior panelling.
w Facelifted –/28 1W 1T.

No.		Livery	Owner	Operator	Depot
9496	r		H	*VT*	TM
9497	r★		H	*MR*	BH
9498	r	**V**	H		KT
9500	r		H	*MR*	BH
9501	w	**FP**	DM		MQ
9502	s	**V**	H		SL
9503	s	**AV**	AW		CF
9504	s	**V**	H	*RV*	CP
9505	s★	**V**	H		MQ
9506	s★	**WX**	H	*MR*	BH
9507	s	**V**	H	*RV*	CP
9508	s	**BG**	CG	*CG*	CP
9509	s	**AV**	AW		CF

AE2F (BSO)   OPEN BRAKE STANDARD

Mark 2F. Air conditioned (Temperature Ltd.). All now refurbished with power-operated vestibule doors, new panels and seat trim. All now further refurbished with carpets and new m.a. sets. –/32 1T. B4 bogies. d. pg. ETH5X.

Lot No. 30861 Derby 1974. 34 t.

No.		Livery	Owner	Operator	Depot
9513		**BP**	FM	*VI*	EM
9516	n	**V**	H		KT
9520	n	**V**	RV	*RV*	CP
9521	★	**AV**	AW	*AW*	CP
9522		**V**	H		KT
9523		**V**	H		KT
9524	n★	**AV**	AW		LM
9525		**WX**	H	*MR*	BH
9526	n★		H	*RV*	CP
9527	n	**V**	RV	*RV*	CP
9529	n	**V**	E		OM
9531		**M**	E		OM
9537	n	**V**	H	*RV*	KT
9538		**V**	H		KT
9539		**AV**	AW		CF

AF2F (DBSO) DRIVING OPEN BRAKE STANDARD

Mark 2F. Air conditioned (Temperature Ltd.). Push & pull (t.d.m. system). Converted from BSO, these vehicles originally had half cabs at the brake end. They have since been refurbished and have had their cabs widened and the cab-end gangways removed. –/30 1W 1T. B4 bogies. d. pg. Cowcatchers. ETH 5X.

9701–9710. Lot No. 30861 Derby 1974. Converted Glasgow 1979. Disc brakes. 34 t.
9711–9713. Lot No. 30861 Derby 1974. Converted Glasgow 1985. 34 t.
9714. Lot No. 30861 Derby 1974. Converted Glasgow 1986. Disc brakes. 34 t.

No.		Livery	Owner	Operator	Depot
9701	(9528)	**AR**	NR		ZA
9702	(9510)	**AR**	NR		ZA
9703	(9517)	**AR**	NR		ZA
9704	(9512)	**AR**	H		LM
9705	(9519)	**AR**	H		KT
9707	(9511)	**AR**	H		KT
9708	(9530)	**AR**	NR		ZA
9709	(9515)	**AR**	H		LM
9710	(9518)	1	H		KT
9711	(9532)	**AR**	H		KT
9713	(9535)	**AR**	H	*RV*	CP
9714	(9536)	**AR**	NR		ZA

AE4E (BUO)   UNCLASSIFIED OPEN BRAKE

Mark 2E. Converted from TSO with new seating for use on Anglo-Scottish overnight services by Railcare, Wolverton. Air conditioned. Stones equipment. B4 bogies. d. –/31 2T. B4 bogies. ETH 4X.

9801–9803. Lot No. 30837 Derby 1972. 33.5 t.
9804–9810. Lot No. 30844 Derby 1972–73. 33.5 t.

9800	(5751)	**FS**	H	*SR*	IS	9806	(5840)	**FS**	H	*SR*	IS
9801	(5760)	**FS**	H	*SR*	IS	9807	(5851)	**FB**	H	*SR*	IS
9802	(5772)	**FS**	H	*SR*	IS	9808	(5871)	**FB**	H	*SR*	IS
9803	(5799)	**FS**	H	*SR*	IS	9809	(5890)	**FS**	H	*SR*	IS
9804	(5826)	**FS**	H	*SR*	IS	9810	(5892)	**FS**	H	*SR*	IS
9805	(5833)	**FS**	H	*SR*	IS						

AJ1G (RFB)   KITCHEN BUFFET FIRST

Mark 3A. Air conditioned. Converted from HST TRFKs, RFBs and FOs. Refurbished with table lamps and burgundy seat trim (except *). 18/– plus two seats for staff use (*24/–). BT10 bogies. d. ETH 14X.

10200–10211. Lot No. 30884 Derby 1977. 39.8 t.
10212–10229. Lot No. 30878 Derby 1975–76. 39.8 t.
10231–10260. Lot No. 30890 Derby 1979. 39.8 t.

Non-standard livery: 10211 EWS dark maroon.

10200	(40519)	* **1**	P	*1*	NC	10232	(10027)	**FP**	P	*GW*	PZ
10202	(40504)	**BG**	CG*CG*		CP	10233	(10013)	**V**	CG		LM
10203	(40506)	* **1**	P	*1*	NC	10235	(10015)	**CD**	CD*CD*		EM
10204	(40505)	**V**	P		LM	10236	(10018)	**V**	WS		LM
10206	(40507)	**V**	P		NC	10237	(10022)	**V**	P		LM
10208	(40517)	**V**	WS		LM	10240	(10003)	**V**	P		LM
10211	(40510)	**0**	E	*E*	TO	10241	(10009)	* **1**	P	*1*	NC
10212	(11049)	**V**	P		WB	10242	(10002)	**V**	P		LM
10213	(11050)	**V**	ST		HT	10245	(10019)	**V**	P		LM
10214	(11034)	* **1**	P	*1*	NC	10246	(10014)	**BG**	CG*CG*		CP
10215	(11032)	**V**	P		LM	10247	(10011)	* **1**	P	*1*	NC
10216	(11041)	* **1**	P	*1*	NC	10249	(10012)	**V**	P		LM
10217	(11051)	**V**	P	*VW*	WB	10250	(10020)	**V**	P		LM
10219	(11047)	**FP**	P	*GW*	PZ	10253	(10026)	**V**	P		LM
10223	(11043)	* **1**	P	*1*	NC	10255	(10010)	**V**	WS		LM
10225	(11014)	**FP**	P	*GW*	PZ	10256	(10028)	**V**	P		GW
10226	(11015)	**V**	P		LM	10257	(10007)	**V**	P		LM
10228	(11035)	* **1**	P	*1*	NC	10259	(10025)	**V**	P		LM
10229	(11059)	* **1**	P	*1*	NC	10260	(10001)	**V**	P		GW
10231	(10016)	**V**	P		LM						

AG2J (RSB)   KITCHEN BUFFET STANDARD

Mark 4. Air conditioned. BT41 bogies. ETH 6X. Rebuilt from first to standard class with bar adjacent to seating area instead of adjacent to end of coach. –/30 1T.

Lot No. 31045 Metro-Cammell 1989–1992. 43.2 t.

10300	**GN**	H	*GN*	BN	10303	**GN**	H	*GN*	BN
10301	**GN**	H	*GN*	BN	10304	**GN**	H	*GN*	BN
10302	**GN**	H	*GN*	BN	10305	**GN**	H	*GN*	BN

10306	**GN**	H *GN*	BN		10320	**GN**	H *GN*	BN
10307	**GN**	H *GN*	BN		10321	**GN**	H *GN*	BN
10308	**GN**	H *GN*	BN		10323	**GN**	H *GN*	BN
10309	**GN**	H *GN*	BN		10324	**GN**	H *GN*	BN
10310	**GN**	H *GN*	BN		10325	**GN**	H *GN*	BN
10311	**GN**	H *GN*	BN		10326	**GN**	H *GN*	BN
10312	**GN**	H *GN*	BN		10328	**GN**	H *GN*	BN
10313	**GN**	H *GN*	BN		10329	**GN**	H *GN*	BN
10315	**GN**	H *GN*	BN		10330	**GN**	H *GN*	BN
10317	**GN**	H *GN*	BN		10331	**GN**	H *GN*	BN
10318	**GN**	H *GN*	BN		10332	**GN**	H *GN*	BN
10319	**GN**	H *GN*	BN		10333	**GN**	H *GN*	BN

AN2G (RMB)
OPEN STANDARD WITH MINIATURE BUFFET

Mark 3A. Air conditioned. Converted from Mark 3 TSOs at Derby 2006. –/52 1T (including 6 Compin Pegasus seats for "priority" use).

Lot No. 30877 Derby 1975–77. 37.8 t.

10401	(12168)	**1**	P *1*	NC		10404	(12068)	**1**	P *1*	NC
10402	(12010)	**1**	P *1*	NC		10405	(12157)	**1**	P *1*	NC
10403	(12135)	**1**	P *1*	NC		10406	(12020)	**1**	P *1*	NC

AU4G (SLEP)   SLEEPING CAR WITH PANTRY

Mark 3A. Air conditioned. Retention toilets. 12 compartments with a fixed lower berth and a hinged upper berth, plus an attendants compartment. 2T BT10 bogies. ETH 7X.

Non-standard livery: 10546 EWS dark maroon.

Lot No. 30960 Derby 1981–83. 41 t.

10501	d	**FS**	P	*SR*	IS		10532	d	**FP**	P	*GW*	PZ
10502	d	**FS**	P	*SR*	IS		10534	d	**FP**	P	*GW*	PZ
10504	d	**FS**	P	*SR*	IS		10542	d	**FS**	P	*SR*	IS
10506	d	**FS**	P	*SR*	IS		10543	d	**FS**	P	*SR*	IS
10507	d	**FS**	P	*SR*	IS		10544	d	**FS**	P	*SR*	IS
10508	d	**FS**	P	*SR*	IS		10546	d	**0**	E	*E*	TO
10513	d	**FS**	P	*SR*	IS		10548	d	**FS**	P	*SR*	IS
10516	d	**FB**	P	*SR*	IS		10551	d	**FS**	P	*SR*	IS
10519	d	**FS**	P	*SR*	IS		10553	d	**FS**	P	*SR*	IS
10520	d	**FS**	P	*SR*	IS		10561	d	**FS**	P	*SR*	IS
10522	d	**FS**	P	*SR*	IS		10562	d	**FB**	P	*SR*	IS
10523	d	**FS**	P	*SR*	IS		10563	d	**FP**	P		LM
10526	d	**FS**	P	*SR*	IS		10565	d	**FS**	P	*SR*	IS
10527	d	**FS**	P	*SR*	IS		10580	d	**FS**	P	*SR*	IS
10529	d	**FS**	P	*SR*	IS		10584	d	**FP**	P	*GW*	PZ
10531	d	**FS**	P	*SR*	IS		10588	d	**BG**	CG *CG*		CP

10589	d	**FP**	P	*GW*	PZ	10605	d	**FS**	P	*SR*	IS
10590	d	**FP**	P	*GW*	PZ	10607	d	**FS**	P	*SR*	IS
10594	d	**FP**	P	*GW*	PZ	10610	d	**FS**	P	*SR*	IS
10596	d		P		KT	10612	d	**FP**	P	*GW*	PZ
10597	d	**FS**	P	*SR*	IS	10613	d	**FS**	P	*SR*	IS
10598	d	**FS**	P	*SR*	IS	10614	d	**FS**	P	*SR*	IS
10600	d	**FS**	P	*SR*	IS	10616	d	**FP**	P	*GW*	PZ
10601	d	**FP**	P	*GW*	PZ	10617	d	**FS**	P	*SR*	IS

AS4G/AQ4G* (SLE/SLED*)   SLEEPING CAR

Mark 3A. Air conditioned. Retention toilets. 13 compartments with a fixed lower berth and a hinged upper berth (* 11 compartments with a fixed lower berth and a hinged upper berth + one compartment for a disabled person). 2T. BT10 bogies. ETH 6X.

Note: 10734 was originally 2914 and used as a Royal Train staff sleeping car. It has 12 berths and a shower room and is ETH11X.

10647–10729. Lot No. 30961 Derby 1980–84. 43.5 t.
10734. Lot No. 31002 Derby/Wolverton 1985. 42.5 t.

10647	d		P		KT	10701	d		P		KT
10648	d*	**FS**	P	*SR*	IS	10703	d	**FS**	P	*SR*	IS
10650	d*	**FS**	P	*SR*	IS	10706	d*	**FS**	P	*SR*	IS
10666	d*	**FS**	P	*SR*	IS	10710	d		CD		KT
10675	d	**FS**	P	*SR*	IS	10714	d*	**FS**	P	*SR*	IS
10680	d*	**FS**	P	*SR*	IS	10718	d*	**FS**	P	*SR*	IS
10683	d	**FS**	P	*SR*	IS	10719	d*	**FS**	P	*SR*	IS
10688	d	**FS**	P	*SR*	IS	10722	d*	**FS**	P	*SR*	IS
10689	d*	**FG**	P	*SR*	IS	10723	d*	**FS**	P	*SR*	IS
10690	d	**FS**	P	*SR*	IS	10729		**VN**	VS	*VS*	CP
10693	d	**FS**	P	*SR*	IS	10734		**VN**	VS	*VS*	CP
10699	d*	**FS**	P	*SR*	IS						

AD1G (FO)   OPEN FIRST

Mark 3A. Air conditioned. All now refurbished with table lamps and new seat cushions and trim. 48/– 2T (* 48/– 1T 1TD). BT10 bogies. d. ETH 6X.

11005–7 were open composites 11905–7 for a time.

Non-standard livery: 11039 EWS dark maroon.

Lot No. 30878 Derby 1975–76. 34.3 t.

11005		**V**	P		LM	11020	**V**	P		LM
11006		**V**	P		LM	11021	**V**	P		NC
11007		**V**	P		LM	11026	**V**	P		LM
11011	*	**V**	P		LM	11027	**V**	P	*VW*	WB
11013		**V**	CD	*CD*	GL	11028	**V**	DR		KM
11016		**V**	P		LM	11029	**CD**	CD	*CD*	GL
11018		**V**	P		LM	11030	**DS**	DR	*DR*	KM
11019		**DS**	DR	*DR*	KM	11031	**BG**	CG	*CG*	CP

11033	**CD**	CD	*CD*	GL	11048	**V**	P	*VW*	WB
11039	**O**	E	*E*	TO	11052	**V**	P		LM
11040	**V**	P		LM	11054	**DS**	DR	*DR*	KM
11042	**V**	P		LM	11058	**V**	P		LM
11044	**DS**	DR	*DR*	KM	11060	**V**	P		LM
11046	**DS**	DR	*DR*	KM					

AD1H (FO)   OPEN FIRST

Mark 3B. Air conditioned. Inter-City 80 seats. All now refurbished with table lamps and new seat cushions and trim. 48/– 2T. BT10 bogies. d. ETH 6X.

† "One" vehicles fitted with disabled toilet and reduced seating including three Compin "Pegasus" seats of the same type as used in standard class (but regarded as first class!). 34/3 1T 1TD 2W.

Lot No. 30982 Derby 1985. 36.5 t.

11064		**V**	P		LM	11083	**BG**	CG	*CG*	CP
11065		**BG**	CG	*CG*	CP	11084	**BG**	CG	*CG*	CP
11066		**1**	P	*1*	NC	11085 †	**1**	P	*1*	NC
11067		**1**	P	*1*	NC	11086	**BG**	CG	*CG*	CP
11068		**1**	P	*1*	NC	11087 †	**1**	P	*1*	NC
11069		**1**	P	*1*	NC	11088 †	**1**	P	*1*	NC
11070		**1**	P	*1*	NC	11089	**BG**	CG	*CG*	CP
11071		**BG**	CG	*CG*	CP	11090 †	**1**	P	*1*	NC
11072		**1**	P	*1*	NC	11091	**1**	P	*1*	NC
11073		**1**	P	*1*	NC	11092 †	**1**	P	*1*	NC
11074		**V**	P		NC	11093 †	**1**	P	*1*	NC
11075		**1**	P	*1*	NC	11094 †	**1**	P	*1*	NC
11076		**1**	P	*1*	NC	11095 †	**1**	P	*1*	NC
11077		**1**	P	*1*	NC	11096 †	**1**	P	*1*	NC
11078 †		**1**	P	*1*	NC	11097	**V**	P		LM
11079		**V**	P	*VW*	WB	11098 †	**1**	P	*1*	NC
11080		**1**	P	*1*	NC	11099 †	**1**	P	*1*	NC
11081		**1**	P	*1*	NC	11100 †	**1**	P	*1*	NC
11082		**1**	P	*1*	NC	11101 †	**1**	P	*1*	NC

AD1J (FO)   OPEN FIRST

Mark 4. Air conditioned. Rebuilt with new interior by Bombardier Wakefield 2003–05 (some converted from standard class vehicles) 46/– 1T. BT41 bogies. ETH 6X.

11201–11273. Lot No. 31046 Metro-Cammell 1989–92. 41.3 t .
11277–11299. Lot No. 31049 Metro-Cammell 1989–92. 41.3 t .

11201	(11201)	**GN**	H	*GN*	BN	11273	(11273)	**GN** H *GN* BN
11219	(11219)	**GN**	H	*GN*	BN	11277	(12408)	**GN** H *GN* BN
11229	(11229)	**GN**	H	*GN*	BN	11278	(12479)	**GN** H *GN* BN
11237	(11237)	**GN**	H	*GN*	BN	11279	(12521)	**GN** H *GN* BN
11241	(11241)	**GN**	H	*GN*	BN	11280	(12523)	**GN** H *GN* BN
11244	(11244)	**GN**	H	*GN*	BN	11281	(12418)	**GN** H *GN* BN

11282	(12524)	**GN** H *GN* BN	11290	(12530)	**GN** H *GN* BN
11283	(12435)	**GN** H *GN* BN	11291	(12535)	**GN** H *GN* BN
11284	(12487)	**GN** H *GN* BN	11292	(12451)	**GN** H *GN* BN
11285	(12537)	**GN** H *GN* BN	11293	(12536)	**GN** H *GN* BN
11286	(12482)	**GN** H *GN* BN	11294	(12529)	**GN** H *GN* BN
11287	(12527)	**GN** H *GN* BN	11295	(12475)	**GN** H *GN* BN
11288	(12517)	**GN** H *GN* BN	11298	(12416)	**GN** H *GN* BN
11289	(12528)	**GN** H *GN* BN	11299	(12532)	**GN** H *GN* BN

AD1J (FOD)   OPEN FIRST (DISABLED)

Mark 4. Air conditioned. Rebuilt from FO by Bombardier Wakefield 2003–05. 42/– 1W 1TD. BT41 bogies. ETH 6X.

Lot No. 31046 Metro-Cammell 1989–92. 40.7 t.

11301	(11215)	**GN** H *GN* BN	11316	(11227)	**GN** H *GN* BN
11302	(11203)	**GN** H *GN* BN	11317	(11223)	**GN** H *GN* BN
11303	(11211)	**GN** H *GN* BN	11318	(11251)	**GN** H *GN* BN
11304	(11257)	**GN** H *GN* BN	11319	(11247)	**GN** H *GN* BN
11305	(11261)	**GN** H *GN* BN	11320	(11255)	**GN** H *GN* BN
11306	(11276)	**GN** H *GN* BN	11321	(11245)	**GN** H *GN* BN
11307	(11217)	**GN** H *GN* BN	11322	(11228)	**GN** H *GN* BN
11308	(11263)	**GN** H *GN* BN	11323	(11235)	**GN** H *GN* BN
11309	(11259)	**GN** H *GN* BN	11324	(11253)	**GN** H *GN* BN
11310	(11272)	**GN** H *GN* BN	11325	(11231)	**GN** H *GN* BN
11311	(11221)	**GN** H *GN* BN	11326	(11206)	**GN** H *GN* BN
11312	(11225)	**GN** H *GN* BN	11327	(11236)	**GN** H *GN* BN
11313	(11210)	**GN** H *GN* BN	11328	(11274)	**GN** H *GN* BN
11314	(11207)	**GN** H *GN* BN	11329	(11243)	**GN** H *GN* BN
11315	(11238)	**GN** H *GN* BN	11330	(11249)	**GN** H *GN* BN

AD1J (FO)   OPEN FIRST

Mark 4. Air conditioned. Rebuilt from FO by Bombardier Wakefield 2003–05. Separate area for 7 smokers, although smoking is no longer allowed. 46/– 1W 1TD. BT41 bogies. ETH 6X.

Lot No. 31046 Metro-Cammell 1989–92. 42.1 t.

11401	(11214)	**GN** H *GN* BN	11414	(11246)	**GN** H *GN* BN
11402	(11216)	**GN** H *GN* BN	11415	(11208)	**GN** H *GN* BN
11403	(11258)	**GN** H *GN* BN	11416	(11254)	**GN** H *GN* BN
11404	(11202)	**GN** H *GN* BN	11417	(11226)	**GN** H *GN* BN
11405	(11204)	**GN** H *GN* BN	11418	(11222)	**GN** H *GN* BN
11406	(11205)	**GN** H *GN* BN	11419	(11250)	**GN** H *GN* BN
11407	(11256)	**GN** H *GN* BN	11420	(11242)	**GN** H *GN* BN
11408	(11218)	**GN** H *GN* BN	11421	(11220)	**GN** H *GN* BN
11409	(11262)	**GN** H *GN* BN	11422	(11232)	**GN** H *GN* BN
11410	(11260)	**GN** H *GN* BN	11423	(11230)	**GN** H *GN* BN
11411	(11240)	**GN** H *GN* BN	11424	(11239)	**GN** H *GN* BN
11412	(11209)	**GN** H *GN* BN	11425	(11234)	**GN** H *GN* BN
11413	(11212)	**GN** H *GN* BN	11426	(11252)	**GN** H *GN* BN

11427 (11200)	**GN** H *GN* BN	11429 (11275)	**GN** H *GN* BN
11428 (11233)	**GN** H *GN* BN	11430 (11248)	**GN** H *GN* BN

AD1J (FO) OPEN FIRST

Mark 4. Air conditioned. Converted from TFRB with new interior by Bombardier Wakefield 2005. 46/– 1T. BT41 bogies. ETH 6X.

Lot No. 31046 Metro-Cammell 1989–92. 41.3 t.

11998 (10314)	**GN** H *GN* BN	11999 (10316)	**GN** H *GN* BN

AC2G (TSO) OPEN STANDARD

Mark 3A. Air conditioned. All refurbished with modified seat backs and new layout and further refurbished with new seat trim. –/76 2T (s –/70 2T 2W, z –/70 1TD 1T 2W). BT10 bogies. d. ETH 6X.

h Coaches modified for "One" with 8 Compin Pegasus seats at saloon ends for "priority" use and a high density layout with more unidirectional seating. –/80 2T.

Note: 12169–72 were converted from open composites 11908–10/22, formerly FOs 11008–10/22.

12004–12167. Lot No. 30877 Derby 1975–77. 34.3 t.
12169–12172. Lot No. 30878 Derby 1975–76. 34.3 t.

12004		**V**	P		LM	12037 h	**1**	P	*1*	NC
12005 h		**1**	P		NC	12038	**BG** CG *CG*	CP
12007		**V**	P		LM	12040 h	**1**	P	*1*	NC
12008		**V**	P		LM	12041 h	**1**	P	*1*	NC
12009 h		**1**	P	*1*	NC	12042 h	**1**	P	*1*	NC
12011		**V**	P	*VW*	WB	12043	**BG** CG *CG*	CP
12012 h		**1**	P	*1*	NC	12045	**V**	P		LM
12013 h		**1**	P	*1*	NC	12046 h	**1**	P	*1*	NC
12014	**BG** CG *CG*	CP				12047 z	**V**	P		LM
12015 h		**1**	P	*1*	NC	12048	**V**	WS		LM
12016		**1**	P	*1*	NC	12049	**1**	P	*1*	NC
12017		**V**	P		LM	12050 s	**V**	P		LM
12019 h		**1**	P	*1*	NC	12051 h	**1**	P	*1*	NC
12021		**1**	P		NC	12052	**V**	P		LM
12022		**V**	P		LM	12053	**BG** CG *CG*	CP
12024 h		**1**	P	*1*	NC	12054 s	**V**	P		LM
12025		**V**	P		LM	12055	**V**	P		LM
12026 h		**1**	P	*1*	NC	12056 h	**1**	P	*1*	NC
12027 h		**1**	P	*1*	NC	12057 h	**1**	P	*1*	NC
12028		**V**	P		LM	12058	**V**	ST		LM
12029		**V**	P		LM	12059 s	**V**	P		LM
12030 h		**1**	P	*1*	NC	12060 h	**1**	P	*1*	NC
12031		**1**	P	*1*	NC	12061 h	**1**	P	*1*	NC
12032 h		**1**	P	*1*	NC	12062 h	**1**	P	*1*	NC
12034		**1**	P	*1*	NC	12063	**V**	P		LM
12035 h		**1**	P	*1*	NC	12064	**1**	P	*1*	NC
12036 s		**V**	P		LM	12065	**V**	P		LM

12066 h	1	P	*1*	NC
12067	1	P	*1*	NC
12069	V	WS		LM
12071	V	P		LM
12072	V	WS		LM
12073 h	1	P	*1*	NC
12075	V	P		LM
12076	V	P		LM
12077	V	P		LM
12078	V	P	*VW*	WB
12079	1	P	*1*	NC
12080	V	P		LM
12081	1	P	*1*	NC
12082 h	1	P	*1*	NC
12083	V	P		LM
12084 h	1	P	*1*	NC
12085 s	V	P		LM
12086 s	V	P		LM
12087 s	V	P		LM
12089	1	P	*1*	NC
12090 h	1	P	*1*	NC
12091 h	1	P	*1*	NC
12092	V	P		LM
12093 h	1	P	*1*	NC
12094	V	P		LM
12095	V	P		LM
12097 h	1	P	*1*	NC
12098	1	P	*1*	NC
12099 h	1	P	*1*	NC
12100 z	FP	P	*GW*	PZ
12101 s	V	P		LM
12102	V	P		LM
12103	1	P	*1*	NC
12104	V	ST		LM
12105 h	1	P	*1*	NC
12106	V	P		LM
12107 h	1	P	*1*	NC
12108	1	P	*1*	NC
12109 h	1	P	*1*	NC
12110 h	1	P	*1*	NC
12111 h	1	P	*1*	NC
12113	V	P		LM
12114 h	1	P	*1*	NC
12115 h	1	P	*1*	NC
12116 h	1	P	*1*	NC
12117	V	WS		LM
12118	1	P	*1*	NC
12119	BG	CG	*CG*	CP
12120 h	1	P	*1*	NC
12122 z	V	P	*VW*	WB
12123	V	P		LM
12124	V	P		LM
12125 h	1	P	*1*	NC
12126 h	1	P	*1*	NC
12127	V	WS		LM
12128 s	V	P		LM
12129 h	1	P	*1*	NC
12130 h	1	P	*1*	NC
12131	V	WS		LM
12132	1	P	*1*	NC
12133	V	P	*VW*	WB
12134	V	P		LM
12137 h	1	P	*1*	NC
12138	V	P	*VW*	WB
12139	V	P		LM
12141	1	P	*1*	NC
12142 z	V	P		LM
12143	1	P	*1*	NC
12144 s	V	P		LM
12145	V	WS		LM
12146	1	P	*1*	NC
12147	1	P	*1*	NC
12148	1	P	*1*	NC
12150 h	1	P	*1*	NC
12151	1	P	*1*	NC
12153	1	P	*1*	NC
12154 h	1	P	*1*	NC
12156	V	P		LM
12158	V	P		LM
12159	1	P	*1*	NC
12160 s	V	P		LM
12161 z	FP	P	*GW*	PZ
12163	V	P		LM
12164	1	P	*1*	NC
12165	V	ST		LM
12166	1	P	*1*	NC
12167 h	1	P	*1*	NC
12169 s	V	WS		LM
12170	1	P	*1*	NC
12171	1	P	*1*	NC
12172 s	V	P		LM

Al2J (TSOE)   OPEN STANDARD (END)

Mark 4. Air conditioned. Rebuilt with new interior by Bombardier Wakefield 2003–05. Separate area for 26 smokers, although smoking is no longer allowed. –/76 1T. BT41 bogies. ETH 6X.

Note: 12232 was converted from the original 12405.

2200–12231. Lot No. 31047 Metro-Cammell 1989–91. 39.5 t.
2232. Lot No. 31049 Metro-Cammell 1989–92. 39.5 t.

2200	**GN**	H *GN*	BN		12217	**GN**	H *GN*	BN
2201	**GN**	H *GN*	BN		12218	**GN**	H *GN*	BN
2202	**GN**	H *GN*	BN		12219	**GN**	H *GN*	BN
2203	**GN**	H *GN*	BN		12220	**GN**	H *GN*	BN
2204	**GN**	H *GN*	BN		12222	**GN**	H *GN*	BN
2205	**GN**	H *GN*	BN		12223	**GN**	H *GN*	BN
2207	**GN**	H *GN*	BN		12224	**GN**	H *GN*	BN
2208	**GN**	H *GN*	BN		12225	**GN**	H *GN*	BN
2209	**GN**	H *GN*	BN		12226	**GN**	H *GN*	BN
2210	**GN**	H *GN*	BN		12227	**GN**	H *GN*	BN
2211	**GN**	H *GN*	BN		12228	**GN**	H *GN*	BN
2212	**GN**	H *GN*	BN		12229	**GN**	H *GN*	BN
2213	**GN**	H *GN*	BN		12230	**GN**	H *GN*	BN
2214	**GN**	H *GN*	BN		12231	**GN**	H *GN*	BN
2215	**GN**	H *GN*	BN		12232	**GN**	H *GN*	BN
2216	**GN**	H *GN*	BN					

AL2J (TSOD)
OPEN STANDARD (DISABLED ACCESS)

Mark 4. Air conditioned. Rebuilt with new interior by Bombardier Wakefield 2003–05. –/68 2W 1TD. BT41 bogies. ETH 6X.

Note: 12331 has been converted from TSO 12531.

12300–12330. Lot No. 31048 Metro-Cammell 1989–91. 39.4 t.
12331. Lot No. 31049 Metro-Cammell 1989–92. 39.4 t.

12300	**GN**	H *GN*	BN		12313	**GN**	H *GN*	BN
12301	**GN**	H *GN*	BN		12315	**GN**	H *GN*	BN
12302	**GN**	H *GN*	BN		12316	**GN**	H *GN*	BN
12303	**GN**	H *GN*	BN		12317	**GN**	H *GN*	BN
12304	**GN**	H *GN*	BN		12318	**GN**	H *GN*	BN
12305	**GN**	H *GN*	BN		12319	**GN**	H *GN*	BN
12307	**GN**	H *GN*	BN		12320	**GN**	H *GN*	BN
12308	**GN**	H *GN*	BN		12321	**GN**	H *GN*	BN
12309	**GN**	H *GN*	BN		12322	**GN**	H *GN*	BN
12310	**GN**	H *GN*	BN		12323	**GN**	H *GN*	BN
12311	**GN**	H *GN*	BN		12324	**GN**	H *GN*	BN
12312	**GN**	H *GN*	BN		12325	**GN**	H *GN*	BN

12326	**GN**	H *GN*	BN		12329	**GN**	H *GN*	BN
12327	**GN**	H *GN*	BN		12330	**GN**	H *GN*	BN
12328	**GN**	H *GN*	BN		12331	**GN**	H *GN*	BN

AC2J (TSO)   OPEN STANDARD

Mark 4. Air conditioned. Rebuilt with new interior by Bombardier Wakefiel 2003–05. –/76 1T. BT41 bogies. ETH 6X.

Lot No. 31049 Metro-Cammell 1989–92. 40.8 t.

Note: 12405 is the second coach to carry that number. It was built from the bodyshell originally intended for 12221. The original 12405 is now 12232.

12400	**GN**	H *GN*	BN		12444	**GN**	H *GN*	BN
12401	**GN**	H *GN*	BN		12445	**GN**	H *GN*	BN
12402	**GN**	H *GN*	BN		12446	**GN**	H *GN*	BN
12403	**GN**	H *GN*	BN		12447	**GN**	H *GN*	BN
12404	**GN**	H *GN*	BN		12448	**GN**	H *GN*	BN
12405	**GN**	H *GN*	BN		12449	**GN**	H *GN*	BN
12406	**GN**	H *GN*	BN		12450	**GN**	H *GN*	BN
12407	**GN**	H *GN*	BN		12452	**GN**	H *GN*	BN
12409	**GN**	H *GN*	BN		12453	**GN**	H *GN*	BN
12410	**GN**	H *GN*	BN		12454	**GN**	H *GN*	BN
12411	**GN**	H *GN*	BN		12455	**GN**	H *GN*	BN
12414	**GN**	H *GN*	BN		12456	**GN**	H *GN*	BN
12415	**GN**	H *GN*	BN		12457	**GN**	H *GN*	BN
12417	**GN**	H *GN*	BN		12458	**GN**	H *GN*	BN
12419	**GN**	H *GN*	BN		12459	**GN**	H *GN*	BN
12420	**GN**	H *GN*	BN		12460	**GN**	H *GN*	BN
12421	**GN**	H *GN*	BN		12461	**GN**	H *GN*	BN
12422	**GN**	H *GN*	BN		12462	**GN**	H *GN*	BN
12423	**GN**	H *GN*	BN		12463	**GN**	H *GN*	BN
12424	**GN**	H *GN*	BN		12464	**GN**	H *GN*	BN
12425	**GN**	H *GN*	BN		12465	**GN**	H *GN*	BN
12426	**GN**	H *GN*	BN		12466	**GN**	H *GN*	BN
12427	**GN**	H *GN*	BN		12467	**GN**	H *GN*	BN
12428	**GN**	H *GN*	BN		12468	**GN**	H *GN*	BN
12429	**GN**	H *GN*	BN		12469	**GN**	H *GN*	BN
12430	**GN**	H *GN*	BN		12470	**GN**	H *GN*	BN
12431	**GN**	H *GN*	BN		12471	**GN**	H *GN*	BN
12432	**GN**	H *GN*	BN		12472	**GN**	H *GN*	BN
12433	**GN**	H *GN*	BN		12473	**GN**	H *GN*	BN
12434	**GN**	H *GN*	BN		12474	**GN**	H *GN*	BN
12436	**GN**	H *GN*	BN		12476	**GN**	H *GN*	BN
12437	**GN**	H *GN*	BN		12477	**GN**	H *GN*	BN
12438	**GN**	H *GN*	BN		12478	**GN**	H *GN*	BN
12439	**GN**	H *GN*	BN		12480	**GN**	H *GN*	BN
12440	**GN**	H *GN*	BN		12481	**GN**	H *GN*	BN
12441	**GN**	H *GN*	BN		12483	**GN**	H *GN*	BN
12442	**GN**	H *GN*	BN		12484	**GN**	H *GN*	BN
12443	**GN**	H *GN*	BN		12485	**GN**	H *GN*	BN

12486	**GN**	H *GN*	BN		12519	**GN**	H *GN*	BN
12488	**GN**	H *GN*	BN		12520	**GN**	H *GN*	BN
12489	**GN**	H *GN*	BN		12522	**GN**	H *GN*	BN
12513	**GN**	H *GN*	BN		12526	**GN**	H *GN*	BN
12514	**GN**	H *GN*	BN		12533	**GN**	H *GN*	BN
12515	**GN**	H *GN*	BN		12534	**GN**	H *GN*	BN
12518	**GN**	H *GN*	BN		12538	**GN**	H *GN*	BN

AA11 (FK)   CORRIDOR FIRST

Mark 1. 42/– 2T. B4 bogies. ETH 3.

Lot No. 30381 Swindon 1959. 33 t.

13229 xk **M** BK *BK* BT | 13230 xk **M** BK *BK* BT

AA1A (FK)   CORRIDOR FIRST

Mark 2A. Pressure ventilated. 42/– 2T. B4 bogies. ETH 4.

Lot No. 30774 Derby 1968. 33 t.

13440 v **G** WC *WC* CS |

AB11 (BFK)   CORRIDOR BRAKE FIRST

Mark 1. 24/– 1T. Commonwealth bogies. ETH 2.

Lot No. 30668 Swindon 1961. 36 t.

Originally numbered in 14xxx series and then renumbered in 17xxx series.

17015 x **G** RV *RV* CP | 17018 v **CH** VT *VT* TM

AB1A (BFK)   CORRIDOR BRAKE FIRST

Mark 2A. Pressure ventilated. 24/– 1T. B4 bogies. ETH 4.

17056–17077. Lot No. 30775 Derby 1967–8. 32 t.
17086–17102. Lot No. 30786 Derby 1968. 32 t.

Originally numbered 14056–102. 17090 was numbered 35503 for a time when declassified.

17056	**M**	E		OM		17090	v **CH**	H		TM
17077	**RV**	RV	*RV*	CP		17102	**M**	WC	*WC*	CS
17086	**RV**	DR		CR						

AX5B   COUCHETTE/GENERATOR COACH

Mark 2B. Formerly part of Royal Train. Converted from a BFK built 1969. Consists of luggage accommodation, guard's compartment, 350 kW diesel generator and staff sleeping accommodation. Pressure ventilated. B5 bogies. ETH 5X.

Non-standard livery: 17105 Oxford blue.

Lot No. 30888 Wolverton 1977. 46 t.

| 17105 | (14105, 2905) | **0** | RV | *RV* | CP |

AB1D (BFK)   CORRIDOR BRAKE FIRST

Mark 2D. Air conditioned (Stones equipment). 24/– 1T. B4 Bogies. ETH 5.

Lot No. 30823 Derby 1971–72. 33.5 t.

Originally numbered 14159–68.

| 17159 | **DS** | DR | *DR* | KM | | 17167 | | **VN** | VS | *VS* | CP |
| 17163 | | VS | | CO | | 17168 | d | **M** | WC | | CS |

AE1H (BFO)   OPEN BRAKE FIRST

Mark 3B. Air conditioned. Fitted with hydraulic handbrake. Refurbished with table lamps and burgundy seat trim. 36/– 1T (w 35/– 1T) BT10 bogies. pg. d. ETH 5X.

Lot No. 30990 Derby 1986. 35.81 t.

| 17173 | w | **FP** | P | *GW* | PZ | | 17175 | w | **FD** | P | *GW* | PZ |
| 17174 | | **FP** | P | *GW* | PZ | | | | | | | |

AA21 (SK)   CORRIDOR STANDARD

Mark 1. Each vehicle has eight compartments. All remaining vehicles have metal window frames and melamine interior panelling. Commonwealth bogies. –/48 2T. ETH 4.

Lot No. 30685 Derby 1961–62. 36 t.

t Rebuilt internally as TSO using components from 4936. –/64 2T.

Originally numbered 25756–25862.

18756	x	**M**	WC	*WC*	CS		18808	x	**M**	WC	*WC*	CS
18767	x	**M**	WC	*WC*	CS		18862	x	**M**	WC	*WC*	CS
18806	xt	**M**	WC	*WC*	CS							

AB31 (BCK)   CORRIDOR BRAKE COMPOSITE

Mark 1. There are two variants depending upon whether the standard class compartments have armrests. Each vehicle has two first class and three standard class compartments. 12/18 2T (12/24 2T *). Commonwealth bogies. ETH 2.

Non-Standard livery: 21269 British racing green & cream lined out in gold.

21241–21246. Lot No. 30669 Swindon 1961–62. Commonwealth bogies. 36 t.
21256. Lot No. 30731 Derby 1963. Commonwealth bogies. 37 t.
21266–21272. Lot No. 30732 Derby 1964. Commonwealth bogies. 37 t.

21241	x	**M**	BK	*BK*	BT		21266	x*	**M**	WC	*WC*	CS
21245	x	**M**	RV	*RV*	CP		21269	*	**GC**	RV	*VS*	SL
21246		**BG**	E		OM		21272	x*	**CH**	RV	*RV*	CP
21256	x	**M**	WC	*WC*	CS							

AB21 (BSK)　　CORRIDOR BRAKE STANDARD

Mark 1. Four compartments. Lot 30721 has metal window frames and melamine interior panelling. –/24 1T. ETH2.

g Fitted with an e.t.s. generator. Weight unknown.

35185. Lot No. 30427 Wolverton 1959. B4 bogies. 33 t.
35452–35469. Lot No. 30721 Wolverton 1963. Commonwealth bogies. 37 t.

35185	x	**M**	BK	*BK*	BT	35459	x	**M**	WC	*WC*	CS
35452	x	**RR**	LW		CP	35469	xg	**M**	E		OM

AB5C　　BRAKE/POWER KITCHEN

Mark 2C. Pressure ventilated. Converted from BFK (declassified to BSK) built 1970. Converted at West Coast Railway Company 2000–01. Consists of 60 kVA generator, guard's compartment and electric kitchen. B5 bogies. ETH 4.

Non-standard livery: British Racing Green with gold lining.

Lot No. 30796 Derby 1969–70. 32.5 t.

35511	(14130, 17130)	**0**	RA	CP

AK51 (RK)　　KITCHEN CAR

Mark 1. Converted 1989/2006 from RBR. Buffet and seating area replaced with additional kitchen and food preparation area. Fluorescent lighting. Commonwealth bogies. ETH 2X.

Lot No. 30628 Pressed Steel 1960–61. 39 t.

Non-standard livery: 80042 Nanking blue.

80041	(1690)	x	**M**	RV	*RV*	CP
80042	(1646)		**0**	FM	*VI*	EM

NZ (DLV)　　DRIVING BRAKE VAN (110 m.p.h.)

Mark 3B. Air conditioned. T4 bogies. dg. ETH 5X.

Lot No. 31042 Derby 1988. 45.18 t.

Non-standard livery: 82146 EWS silver.

82101	**V**	P	*VW*	WB	82111	**V**	P		LM
82102	**1**	P	*1*	NC	82112	**1**	P	*1*	NC
82103	**1**	P	*1*	NC	82113	**V**	P		LM
82104	**V**	P		NC	82114	**1**	P	*1*	NC
82105	**1**	P	*1*	NC	82115	**V**	P		LM
82106	**V**	P		ZB	82116	**V**	P		LM
82107	**1**	P	*1*	NC	82117	**V**	P		LM
82108	**V**	P		LM	82118	**1**	P	*1*	NC
82109	**V**	P		ZB	82120	**V**	P		LM
82110	**V**	P		LM	82121	**1**	P	*1*	NC

82122	V	P		LM
82123	V	P		LM
82124	V	P		LM
82125	V	P		LM
82126	V	P	VW	WB
82127	1	P	1	NC
82128	V	P		LM
82129	V	P		LM
82130	V	P		LM
82131	V	P		LM
82132	1	P	1	NC
82133	1	P	1	NC
82134	V	P		LM
82135	V	P		LM
82136	1	P	1	NC
82137	V	P		LM
82138	V	P		LM
82139	1	P	1	NC
82140	V	P		LM
82141	V	P		LM
82142	V	P		LM
82143	1	P	1	NC
82144	V	P		LM
82145	V	P		CE
82146	0	E	E	TO
82147	V	P		LM
82148	V	P		LM
82149	V	P		LM
82150	V	P		LM
82151	V	P		CE
82152	1	P	1	NC

NZ (DLV)   DRIVING BRAKE VAN (140 m.p.h.)

Mark 4. Air conditioned. Swiss-built (SIG) bogies. dg. ETH 6X.

Fitted with transceiver "domes" for wi-fi.

Lot No. 31043 Metro-Cammell 1988. 45.18 t.

82200	GN	H	GN	BN
82201	GN	H	GN	BN
82202	GN	H	GN	BN
82203	GN	H	GN	BN
82204	GN	H	GN	BN
82205	GN	H	GN	BN
82206	GN	H	GN	BN
82207	GN	H	GN	BN
82208	GN	H	GN	BN
82209	GN	H	GN	BN
82210	GN	H	GN	BN
82211	GN	H	GN	BN
82212	GN	H	GN	BN
82213	GN	H	GN	BN
82214	GN	H	GN	BN
82215	GN	H	GN	BN
82216	GN	H	GN	BN
82217	GN	H	GN	BN
82218	GN	H	GN	BN
82219	GN	H	GN	BN
82220	GN	H	GN	BN
82222	GN	H	GN	BN
82223	GN	H	GN	BN
82224	GN	H	GN	BN
82225	GN	H	GN	BN
82226	GN	H	GN	BN
82227	GN	H	GN	BN
82228	GN	H	GN	BN
82229	GN	H	GN	BN
82230	GN	H	GN	BN
82231	GN	H	GN	BN

NAMED COACHES

The following miscellaneous coaches carry names:

1200	AMBER	3356	TENNYSON
1659	CAMELOT	3364	SHAKESPEARE
1800	TINTAGEL	3384	DICKENS
3105	JULIA	3390	CONSTABLE
3113	JESSICA	3397	WORDSWORTH
3117	CHRISTINA	3426	ELGAR
3128	VICTORIA	5193	CLAN MACLEOD
3130	PAMELA	5212	CAPERKAILZIE
3136	DIANA	5229	THE GREEN KNIGHT
3143	PATRICIA	5239	THE RED KNIGHT
3174	GLAMIS	5278	MELISANDE
3182	WARWICK	5365	Deborah
3188	SOVEREIGN	5376	Michaela
3223	DIAMOND	5419	SIR LANCELOT
3228	AMETHYST	9391	PENDRAGON
3229	JADE	10569	LEVIATHAN
3231	Apollo	10729	CREWE
3240	SAPPHIRE	10734	BALMORAL
3244	EMERALD	17013	ALBANNACH SGIATHACH
3247	CHATSWORTH	17086	Georgina
3267	BELVOIR	35518	MERLIN
3273	ALNWICK	82126	Wembley Traincare Centre
3275	HARLECH	82217	OFF TO THE RACES
3330	BRUNEL	82219	Duke of Edinburgh
3348	GAINSBOROUGH		

2. HIGH SPEED TRAIN TRAILER CARS

HSTs consist of a number of trailer cars (usually seven to nine) with a power car at each end. All trailer cars are classified Mark 3 and have BT10 bogies with disc brakes and central door locking. Heating is by a 415 V three-phase supply and vehicles have air conditioning. Maximum speed is 125 m.p.h.

All vehicles underwent a mid-life refurbishment in the 1980s, and a further refurbishment programme was completed in November 2000, with each train operating company having a different scheme as follows:

First Great Western. Green seat covers and extra partitions between seat bays.

Great North Eastern Railway. New ceiling lighting panels and brown seat covers. First class vehicles have table lamps and imitation walnut plastic end panels.

Virgin Cross-Country. Green seat covers. Standard class vehicles had four seats in the centre of each carriage replaced with a luggage stack. All have now passed to other operators or are in store.

Midland Mainline. Grey seat covers, redesigned seat squabs, side carpeting and two seats in the centre of each standard class carriage and one in first class carriages replaced with a luggage stack.

Since then the remaining three operators of HSTs have embarked on separate, and very different, refurbishment projects:

Midland Mainline were first to refurbish their vehicles a second time during 2003/04. This involved fitting new fluorescent and halogen ceiling lighting, although the original seats were retained in first and standard class, but with blue upholstery.

First Great Western started a major rebuild of their HST sets in late 2006, with the first set entering traffic in early 2007. By spring 2008 the company plans to have 53 sets in traffic with new interiors featuring new lighting and seating throughout. First class seats, with leather upholstery, are made by Primarius UK and standard class seats are made by Grammer. Some sets are being reduced to 7-cars (without a buffet car) with "high density" seating layouts in standard class, with almost all seats arranged in a unidirectional layout. The refurbishment work is being carried out at Derby and Ilford Works by Bombardier.

GNER modernised their buffet cars with new corner bars in 2004 and at the same time each HST set was made up to 9-cars with an extra standard class vehicle added with a disabled person's toilet.

At the end of 2006 GNER embarked on a major rebuild of its HST sets, with the work being carried out at Wabtec, Doncaster. All vehicles will have similar interiors to the Mark 4 "Mallard" fleet, with new Primarius seats throughout. 13 sets are to be refurbished and the work is due for completion by March 2009.

Ten sets ex-Virgin Cross-Country, and some spare vehicles, were temporarily allocated to Midland Mainline for the interim service to Manchester during 2003/04 and had a facelift. Buffet cars were converted from TRSB to TRFB and renumbered in the 408xx series. These sets are now in use with First Great Western or GNER (although those in use with First Great Western are due to come off lease).

Open access operator **Grand Central** is forming up three 6-car sets made up of 2TF–TRSB–3TS with the TF and TS being converted from loco-hauled Mark 3 stock. They also bought three TGSs, but are not now planning to use them. They are due to launch a Sunderland–King's Cross service by the end of 2006.

Arriva Cross-Country plan to reintroduce HST sets to the Cross-Country network on a regular basis. Five sets will be introduced during 2008/09, with some of these also using converted loco-hauled Mark 3 stock.

TOPS Type Codes

TOPS type codes for HST trailer cars are made up as follows:

(1) Two letters denoting the layout of the vehicle as follows:

GH	Open	GL	Kitchen
GJ	Open with Guard's compartment	GN	Buffet

(2) A digit for the class of passenger accommodation

1	First	2	Standard (formerly second).

(3) A suffix relating to the build of coach.

G   Mark 3

Operator Codes

The normal operator codes are given in brackets after the TOPS codes. These are as follows:

TF	Trailer First	TGS	Trailer Guard's Standard
TRB	Trailer Buffet First	TRSB	Trailer Buffet Standard
TRFB	Trailer Buffet First	TS	Trailer Standard

General Note: + Ex-store vehicles assigned for lease to First Great Western after refurbishment.

GN1G (TRFB)   TRAILER BUFFET FIRST

Converted from TRSB by fitting first class seats. Renumbered from 404xx series by subtracting 200. 23/–.

40204–40228. Lot No. 30883 Derby 1976–77. 36.12 t.
40231. Lot No. 30899 Derby 1978–79. 36.12 t.

* Refurbished First Great Western vehicles. New Primarius leather seats.

40204	*	**FD**	A	*GW*	LA	40210	*	**FD**	A	*GW*	LA
40205	*	**FD**	A	*GW*	LA	40221	*	**FD**	A	*GW*	LA
40207	*	**FD**	A	*GW*	LA	40228		**FG**	A	*GW*	LA
40208		**FG**	A	*GW*	LA	40231	*	**FD**	A	*GW*	LA
40209		**FG**	A	*GW*	LA						

GK2G (TRSB)   TRAILER BUFFET STANDARD

Renumbered from 400xx series by adding 400. –/33 1W.

40402–40426. Lot No. 30883 Derby 1976–77. 36.12 t.
40433/40434. Lot No. 30899 Derby 1978–79. 36.12 t.

Note: 40433/40434 were numbered 40233/40234 for a time when fitted with 23 first class seats.

40402	**V**	P	LM	40424	**GC**	ST	HT
40403	**V**	P	LM	40426	**GC**	ST	HT
40416	**V**	P	LM	40433	**GC**	ST	HT
40419	**V**	P	LM	40434	**V**	P	LM

GK1G (TRFB)   TRAILER BUFFET FIRST

These vehicles have larger kitchens than the 402xx and 404xx series vehicles, and are used in trains where full meal service is required. They were renumbered from the 403xx series (in which the seats were unclassified) by adding 400 to the previous number. 17/–.

40700–40721. Lot No. 30921 Derby 1978–79. 38.16 t.
40722–40735. Lot No. 30940 Derby 1979–80. 38.16 t.
40736–40753. Lot No. 30948 Derby 1980–81. 38.16 t.
40754–40757. Lot No. 30966 Derby 1982. 38.16 t.

* Refurbished First Great Western vehicles. New Primarius leather seats.
m Refurbished GNER "Mallard" vehicles with new Primarius seating.
† Fitted with transceiver "dome" for wi-fi.

40700		**MN**	P	*MM*	NL	40723	**MN**	A	*MM*	NL
40701	†	**GN**	P	*GN*	EC	40724	**FG**	A	*GW*	LA
40702		**MN**	P	*MM*	NL	40725	**FG**	A	*GW*	LA
40703	*	**FD**	A	*GW*	LA	40726	**FG**	A	*GW*	LA
40704	†	**GN**	A	*GN*	EC	40727 *	**FD**	A	*GW*	LA
40705	†	**GN**	A	*GN*	EC	40728	**MN**	P	*MM*	NL
40706	†	**GN**	A	*GN*	EC	40729	**MN**	P	*MM*	NL
40707	*	**FD**	A	*GW*	LA	40730	**MN**	P	*MM*	NL
40708	†	**GN**	P	*GN*	EC	40731	**FG**	A	*GW*	LA
40709		**FG**	A	*GW*	LA	40732	**MN**	A	*MM*	NL
40710	*	**FD**	A	*GW*	LA	40733 *	**FD**	A	*GW*	LA
40711	†	**GN**	A	*GN*	EC	40734 *	**FD**	A	*GW*	LA
40712		**FG**	A	*GW*	LA	40735 m†	**GN**	A	*GN*	EC
40713	*	**FD**	A	*GW*	LA	40736	**FG**	A	*GW*	LA
40714		**FG**	A	*GW*	LA	40737 †	**GN**	A	*GN*	EC
40715	*	**FD**	A	*GW*	LA	40738	**FG**	A	*GW*	LA
40716	*	**FD**	A	*GW*	LA	40739 *	**FD**	A	*GW*	LA
40717		**FG**	A	*GW*	LA	40740 †	**GN**	A	*GN*	EC
40718	*	**FD**	A	*GW*	LA	40741	**MN**	P	*MM*	NL
40720	†	**GN**	A	*GN*	EC	40742 †	**GN**	A	*GN*	EC
40721	*	**FD**	A	*GW*	LA	40743 *	**FD**	A	*GW*	LA
40722	*	**FD**	A	*GW*	LA	40744	**FG**	A	*GW*	LA

40745		FG	A	GW	LA
40746		MN	P	MM	NL
40747		FG	A	GW	LA
40748	†	GN	A	GN	EC
40749		MN	P	MM	NL
40750	m†	GN	A	GN	EC
40751		MN	P	MM	NL

40752	*	FD	A	GW	LA
40753		MN	P	MM	NL
40754		MN	P	MM	NL
40755	*	FD	A	GW	LA
40756		MN	P	MM	NL
40757	*	FD	A	GW	LA

GL1G (TRFB)   TRAILER BUFFET FIRST

These vehicles have been converted from TRSBs in the 404xx series to be similar to the 407xx series vehicles. 17/–.

40801–40803/40805/40808/40809/40811. Lot No. 30883 Derby 1976–77. 38.16 t.
40804/40806/40807/40810. Lot No. 30899 Derby 1978–79. 38.16 t.

Note: 40802/40804/40811 were numbered 40212/40232/40211 for a time when fitted with 23 first class seats.
m Refurbished GNER "Mallard" vehicles with new Primarius seating.

40801	(40027, 40427)		FG	P	GW	LA
40802	(40012, 40412)		FG	P	GW	LA
40803	(40018, 40418)		FG	P	GW	LA
40804	(40032, 40432)	†	GN	P	GN	EC
40805	(40020, 40420)	m†	GN	P	GN	EC
40806	(40029, 40429)		FG	P	GW	LA
40807	(40035, 40435)		FG	P	GW	LA
40808	(40015, 40415)		FG	P	GW	LA
40809	(40014, 40414)		FG	P	GW	LA
40810	(40030, 40430)		FG	P	GW	LA
40811	(40011, 40411)		GN	P	GN	EC

GN1G (TRB)   TRAILER BUFFET FIRST

Vehicles owned by First Group. Converted from TRSB by First Great Western. 23/–.

40900/40902/40904. Lot No. 30883 Derby 1976–77. 36.12 t.
40901/40903. Lot No. 30899 Derby 1978–79. 36.12 t.

* Refurbished First Great Western vehicles. New Primarius leather seats.

40900	(40022, 40422)	*	FD	FG	GW	LA
40901	(40036, 40436)	*	FD	FG	GW	LA
40902	(40023, 40423)	*	FD	FG	GW	LA
40903	(40037, 40437)	*	FD	FG	GW	LA
40904	(40001, 40401)	*	FD	FG	GW	LA

GH1G (TF)   TRAILER FIRST

41003–41056. Lot No. 30881 Derby 1976–77. 33.66 t.
41057–41120. Lot No. 30896 Derby 1977–78. 33.66 t.
41121–41148. Lot No. 30938 Derby 1979–80. 33.66 t.
41149–41166. Lot No. 30947 Derby 1980. 33.66 t.
41167–41169. Lot No. 30963 Derby 1982. 33.66 t.

41170. Lot No. 30967 Derby 1982. Former prototype vehicle. 33.66 t.
41179/41180. Lot No. 30884 Derby 1976–77. 33.66 t.
41181–41184/41189. Lot No. 30939 Derby 1979–80. 33.66 t.
41185–41188. Lot No. 30969 Derby 1982. 33.66 t.

As built and m 48/– 2T.
*   Refurbished First Great Western vehicles. New Primarius leather seats.
s   Fitted with centre luggage stack, disabled toilet and wheelchair space.
     46/– 1T 1TD 1W.
t   Fitted with disabled toilet and wheelchair space. 47/– 1T 1TD 1W.
w   Wheelchair space. 47/– 2T 1W.
x   Toilet removed for trolley space. 48/– 1T.

No.		Op.		Liv.	Depot
41003	*x	**FD**	A	*GW*	LA
41004	*x	**FD**	A	*GW*	LA
41005	*	**FD**	A	*GW*	LA
41006	*	**FD**	A	*GW*	LA
41007	*	**FD**	A	*GW*	LA
41008	*	**FD**	A	*GW*	LA
41009		**FG**	A	*GW*	LA
41010		**FG**	A	*GW*	LA
41011	*	**FD**	A	*GW*	LA
41012	*	**FD**	A	*GW*	LA
41015		**FG**	A	*GW*	LA
41016	w	**FG**	A	*GW*	LA
41017		**FG**	A	*GW*	LA
41018		**FG**	A	*GW*	LA
41019		**FG**	A	*GW*	LA
41020		**FG**	A	*GW*	LA
41021	*	**FD**	A	*GW*	LA
41022	*	**FD**	A	*GW*	LA
41023		**FG**	A	*GW*	LA
41024		**FG**	A	*GW*	LA
41025	t	**MN**	A	*MM*	NL
41026		**MN**	A	*MM*	NL
41027		**FG**	A	*GW*	LA
41028		**FG**	A	*GW*	LA
41029		**FG**	A	*GW*	LA
41030		**FG**	A	*GW*	LA
41031	*x	**FD**	A	*GW*	LA
41032	*w	**FD**	A	*GW*	LA
41033	*x	**FD**	A	*GW*	LA
41034	*w	**FD**	A	*GW*	LA
41035		**MN**	A	*MM*	NL
41036	t	**MN**	A	*MM*	NL
41037	*x	**FD**	A	*GW*	LA
41038	*w	**FD**	A	*GW*	LA
41039	m	**GN**	A	*GN*	EC
41040	mw	**GN**	A	*GN*	EC
41041	s	**MN**	P	*MM*	NL
41044	w	**GN**	A	*GN*	EC
41045	*x	**FD**	FG	*GW*	LA
41046	s	**MN**	P	*MM*	NL
41051	*x	**FD**	A	*GW*	LA
41052	*w	**FD**	A	*GW*	LA
41055	*	**FD**	A	*GW*	LA
41056	*	**FD**	A	*GW*	LA
41057		**MN**	P	*MM*	NL
41058	mw	**GN**	P	*GN*	EC
41059	*w	**FD**	FG	*GW*	LA
41061		**MN**	P	*MM*	NL
41062	w	**MN**	P	*MM*	NL
41063		**MN**	P	*MM*	NL
41064	s	**MN**	P	*MM*	NL
41065	*	**FD**	A	*GW*	LA
41066		**GN**	P	*GN*	EC
41067	s	**MN**	P	*MM*	NL
41068	s	**MN**	P	*MM*	NL
41069	s	**MN**	P	*MM*	NL
41070	s	**MN**	P	*MM*	NL
41071		**MN**	P	*MM*	NL
41072	s	**MN**	P	*MM*	NL
41075		**MN**	P	*GW*	LA
41076	s	**MN**	P	*GW*	LA
41077		**MN**	P	*MM*	NL
41078		**MN**	P	*MM*	NL
41079		**MN**	P	*MM*	NL
41080		**GN**	P	*GN*	EC
41081	*x	**FD**	P	*GW*	LA
41083		**GN**	P	*GN*	EC
41084	s	**MN**	P	*MM*	NL
41085	*x	**FD**	FG	*GW*	LA
41086	*x	**FD**	FG	*GW*	LA
41087		**GN**	A	*GN*	EC
41088	w	**GN**	A	*GN*	EC
41089	*	**FD**	A	*GW*	LA
41090	w	**GN**	A	*GN*	EC
41091		**GN**	A	*GN*	EC
41092	w	**GN**	A	*GN*	EC
41093	*x	**FD**	A	*GW*	LA
41094	*w	**FD**	A	*GW*	LA

No.					
41095	w	**GN**	P	*GN*	EC
41096	*x	**FD**	P	*GW*	LA
41097	m	**GN**	A	*GN*	EC
41098	w	**GN**	A	*GN*	EC
41099		**GN**	A	*GN*	EC
41100	w	**GN**	A	*GN*	EC
41101		**FG**	A	*GW*	LA
41102		**FG**	A	*GW*	LA
41103		**FG**	A	*GW*	LA
41104		**FG**	A	*GW*	LA
41105		**FG**	A	*GW*	LA
41106		**FG**	A	*GW*	LA
41107	w	**FG**	P	*GW*	LA
41108	*w	**FD**	P	*GW*	LA
41109	*x	**FD**	P	*GW*	LA
41110	*w	**FD**	A	*GW*	LA
41111		**MN**	P	*MM*	NL
41112		**MN**	P	*MM*	NL
41113	s	**MN**	P	*MM*	NL
41114	*x	**FD**	FG	*GW*	LA
41115	m	**GN**	P	*GN*	EC
41116		**FG**	A	*GW*	LA
41117		**MN**	P	*MM*	NL
41118	w	**GN**	A	*GN*	EC
41119	*x	**FD**	P	*GW*	LA
41120		**GN**	A	*GN*	EC
41121	*x	**FD**	A	*GW*	LA
41122	*w	**FD**	A	*GW*	LA
41123		**FG**	A	*GW*	LA
41124		**FG**	A	*GW*	LA
41125	*	**FD**	A	*GW*	LA
41126	*	**FD**	A	*GW*	LA
41127		**FG**	A	*GW*	LA
41128		**FG**	A	*GW*	LA
41129	*	**FD**	A	*GW*	LA
41130	*	**FD**	A	*GW*	LA
41131	*	**FD**	A	*GW*	LA
41132	*	**FD**	A	*GW*	LA
41133	*	**FD**	A	*GW*	LA
41134	*	**FD**	A	*GW*	LA
41135		**FG**	A	*GW*	LA
41136	*w	**FD**	A	*GW*	LA
41137		**FG**	A	*GW*	LA
41138		**FG**	A	*GW*	LA
41139	*	**FD**	A	*GW*	LA
41140	*	**FD**	A	*GW*	LA
41141	*x	**FD**	A	*GW*	LA
41142	*w	**FD**	A	*GW*	LA
41143		**FG**	A	*GW*	LA
41144		**FG**	A	*GW*	LA
41145	*x	**FD**	A	*GW*	LA
41146	*w	**FD**	A	*GW*	LA
41147	w	**FG**	P	*GW*	LA
41148	*x	**FD**	P	*GW*	LA
41149	*w	**FD**	P	*GW*	LA
41150	w	**GN**	A	*GN*	EC
41151		**GN**	A	*GN*	EC
41152		**GN**	A	*GN*	EC
41153		**GN**	P	*GN*	EC
41154	s	**MN**	P	*MM*	NL
41155		**MN**	P	*MM*	NL
41156		**MN**	P	*MM*	NL
41157	*x	**FD**	A	*GW*	LA
41158	*w	**FD**	A	*GW*	LA
41159	w	**GN**	P	*GN*	EC
41160	*w	**FD**	FG	*GW*	LA
41161	*w	**FD**	P	*GW*	LA
41162	*w	**FD**	FG	*GW*	LA
41163	*x	**FD**	FG	*GW*	LA
41164	w	**GN**	A	*GN*	EC
41165	mw	**GN**	P	*GN*	EC
41166	*w	**FD**	FG	*GW*	LA
41167	*w	**FD**	FG	*GW*	LA
41168	*x	**FD**	P	*GW*	LA
41169	*w	**FD**	P	*GW*	LA

No.						
41170	(41001)		**GN**	A	*GN*	EC
41179	(40505)		**FG**	A	*GW*	LA
41180	(40511)		**FG**	A	*GW*	LA
41181	(42282)	*x	**FD**	P	*GW*	LA
41182	(42278)	*w	**FD**	P	*GW*	LA
41183	(42274)	*w	**FD**	P	*GW*	LA
41184	(42270)	*x	**FD**	P	*GW*	LA
41185	(42313)		**GN**	P	*GN*	EC
41186	(42312)		**FG**	P	*GW*	LA
41187	(42311)	*w	**FD**	P	*GW*	LA
41188	(42310)		**FG**	P	*GW*	LA
41189	(42298)	*w	**FD**	P	*GW*	LA
41190	(42088)		**V**	P		LM
41191	(42318)	*	**FD**	P	*GW*	LA
41192	(42246)	*	**FD**	P	*GW*	LA

The following vehicles are converted from loco-hauled Mark 3 vehicles.

41201 (11045)	**GC** ST	HT	41204 (11023)	**GC** ST	HT
41202 (11017)	**GC** ST	HT	41205 (11036)	**GC** ST	HT
41203 (11038)	**GC** ST	HT	41206 (11055)	**GC** ST	HT

GH2G (TS)   TRAILER STANDARD

42003–42090/42362. Lot No. 30882 Derby 1976–77. 33.60 t.
42091–42250. Lot No. 30897 Derby 1977–79. 33.60 t.
42251–42305. Lot No. 30939 Derby 1979–80. 33.60 t.
42306–42322. Lot No. 30969 Derby 1982. 33.60 t.
42323–42341. Lot No. 30983 Derby 1984–85. 33.60 t.
42342/42360. Lot No. 30949 Derby 1982. 33.47 t. Converted from TGS.
42343/42345. Lot No. 30970 Derby 1982. 33.47 t. Converted from TGS.
42344/42361. Lot No. 30964 Derby 1982. 33.47 t. Converted from TGS.
42346/42347/42350/42351. Lot No. 30881 Derby 1976–77. 33.66 t. Converted from TF.
42348/42349/42363. Lot No. 30896 Derby 1977–78. 33.66 t. Converted from TF.
42352/42354. Lot No. 30897 Derby 1977. Were TF from 1983 to 1992. 33.66 t.
42353/42355–42357. Lot No. 30967 Derby 1982. Ex-prototype vehicles. 33.66 t.

Standard seating and m –/76 2T.
* Refurbished First Great Western vehicles. New Grammer seating. –/80 2T (unless h – high density).
m Refurbished GNER "Mallard" vehicles with new Primarius seating.
d FGW vehicles with disabled persons toilet and 5 tip-up seats. –/65 1T 1TD 2W. (–/68 1T 1TD 2W d).
h "High density" FGW vehicles. –/84 2T.
k "High density" FGW refurbished vehicle with disabled persons toilet and 6 tip-up seats. –/72 1T 1TD 2W.
s Centre luggage stack –/72 2T.
u Centre luggage stack –/74 2T.
w Centre luggage stack and wheelchair space –72 2T 1W.
x Unrefurbished vehicles (ex-Virgin Cross-Country). Seats removed for wheelchair spaces. –/68 2T 2W.
† Disabled persons toilet (GNER) –/62 1T 1TD 1W.

42158 was numbered 41177 for a time when fitted with first class seats.

42003		**FG**	A	*GW*	LA	42021 d	**FG**	A	*GW*	LA
42004	*d	**FD**	A	*GW*	LA	42023	**FG**	A	*GW*	LA
42005	*	**FD**	A	*GW*	LA	42024 d	**FG**	A	*GW*	LA
42006		**FG**	A	*GW*	LA	42025	**FG**	A	*GW*	LA
42007 d		**FG**	A	*GW*	LA	42026	**FG**	A	*GW*	LA
42008 d		**FG**	A	*GW*	LA	42027	**FG**	A	*GW*	LA
42009		**FG**	A	*GW*	LA	42028	**FG**	A	*GW*	LA
42010 *		**FD**	A	*GW*	LA	42029	**FG**	A	*GW*	LA
42012 d		**FG**	A	*GW*	LA	42030	*d	**FD**	A	*GW*	LA
42013		**FG**	A	*GW*	LA	42031	*	**FD**	A	*GW*	LA
42014		**FG**	A	*GW*	LA	42032	*	**FD**	A	*GW*	LA
42015	*d	**FD**	A	*GW*	LA	42033	**FG**	A	*GW*	LA
42016	*	**FD**	A	*GW*	LA	42034	**FG**	A	*GW*	LA
42019	*	**FD**	A	*GW*	LA	42035	**FG**	A	*GW*	LA

42036	u	**MN**	A	*MM*	NL	42089	*h	**FD**	A	*GW*	LA
42037	u	**MN**	A	*MM*	NL	42090	*h	**FD**	P	*GW*	LA

No.	flag	code	c1	code2	reg
42036	u	**MN**	A	*MM*	NL
42037	u	**MN**	A	*MM*	NL
42038	u	**MN**	A	*MM*	NL
42039		**FG**	A	*GW*	LA
42040		**FG**	A	*GW*	LA
42041		**FG**	A	*GW*	LA
42042		**FG**	A	*GW*	LA
42043	*	**FD**	A	*GW*	LA
42044		**FG**	A	*GW*	LA
42045	*	**FD**	A	*GW*	LA
42046	*	**FD**	A	*GW*	LA
42047	*	**FD**	A	*GW*	LA
42048	*h	**FD**	A	*GW*	LA
42049	*h	**FD**	A	*GW*	LA
42050	*h	**FD**	A	*GW*	LA
42051	u	**MN**	A	*MM*	NL
42052	u	**MN**	A	*MM*	NL
42053	u	**MN**	A	*MM*	NL
42054		**FG**	A	*GW*	LA
42055	*	**FD**	A	*GW*	LA
42056	*	**FD**	A	*GW*	LA
42057	m	**GN**	A	*GN*	EC
42058	m	**GN**	A	*GN*	EC
42059	m	**GN**	A	*GN*	EC
42060		**FG**	A	*GW*	LA
42061		**FG**	A	*GW*	LA
42062	*k	**FD**	A	*GW*	LA
42063		**GN**	A	*GN*	EC
42064		**GN**	A	*GN*	EC
42065		**GN**	A	*GN*	EC
42066	*k	**FD**	A	*GW*	LA
42067	*h	**FD**	A	*GW*	LA
42068	*h	**FD**	A	*GW*	LA
42069	*d	**FD**	A	*GW*	LA
42070		**FG**	A	*GW*	LA
42071	*	**FD**	A	*GW*	LA
42072	*	**FD**	A	*GW*	LA
42073		**FG**	A	*GW*	LA
42074		**FG**	A	*GW*	LA
42075	*	**FD**	A	*GW*	LA
42076	*	**FD**	A	*GW*	LA
42077	*	**FD**	A	*GW*	LA
42078	*	**FD**	A	*GW*	LA
42079	*	**FD**	A	*GW*	LA
42080	*	**FD**	A	*GW*	LA
42081	*d	**FD**	A	*GW*	LA
42083	*	**FD**	A	*GW*	LA
42084	*d	**FD**	P	*GW*	LA
42085	*h	**FD**	P	*GW*	LA
42086	*h	**FD**	P	*GW*	LA
42087	*h	**FD**	P	*GW*	LA
42089	*h	**FD**	A	*GW*	LA
42090	*h	**FD**	P	*GW*	LA
42091	†	**GN**	A	*GN*	EC
42092		**V**	FG		ZI+
42093	*h	**FD**	FG	*GW*	LA
42094	*h	**FD**	FG	*GW*	LA
42095		**FG**	FG	*GW*	LA
42096		**FG**	A	*GW*	LA
42097	w	**MN**	A	*MM*	NL
42098		**FG**	A	*GW*	LA
42099	*	**FD**	A	*GW*	LA
42100	u	**MN**	P	*MM*	NL
42101		**GN**	P	*GN*	EC
42102		**GN**	P	*GN*	EC
42103		**GN**	FG		ZI+
42104		**GN**	A	*GN*	EC
42105		**V**	FG		ZI+
42106		**GN**	A	*GN*	EC
42107	*	**FD**	A	*GW*	LA
42108	*h	**FD**	FG	*GW*	LA
42109	m	**GN**	P	*GN*	EC
42110	m	**GN**	P	*GN*	EC
42111	u	**MN**	P	*MM*	NL
42112	u	**MN**	P	*MM*	NL
42113	u	**MN**	P	*MM*	NL
42115	*d	**FD**	P	*GW*	LA
42116	†	**GN**	A	*GN*	EC
42117	m	**GN**	P	*GN*	EC
42118	*	**FD**	A	*GW*	LA
42119	u	**MN**	P	*MM*	NL
42120	u	**MN**	P	*MM*	NL
42121	u	**MN**	P	*MM*	NL
42122		**GN**	A	*GN*	EC
42123	u	**MN**	P	*MM*	NL
42124	u	**MN**	P	*MM*	NL
42125	u	**MN**	P	*MM*	NL
42126	*	**FD**	A	*GW*	LA
42127	†	**GN**	A	*GN*	EC
42128	†	**GN**	A	*GN*	EC
42129	*	**FD**	A	*GW*	LA
42130		**V**	P		ZB
42131	u	**MN**	P	*MM*	NL
42132	u	**MN**	P	*MM*	NL
42133	u	**MN**	P	*MM*	NL
42134		**GN**	A	*GN*	EC
42135	u	**MN**	P	*MM*	NL
42136	u	**MN**	P	*MM*	NL
42137	u	**MN**	P	*MM*	NL
42138	*k	**FD**	A	*GW*	LA
42139	u	**MN**	P	*MM*	NL
42140	u	**MN**	P	*MM*	NL

42141	u	**MN**	P	*MM*	NL	42193	m	**GN**	A	*GN*	EC
42143	*	**FD**	A	*GW*	LA	42194	w	**MN**	P	*MM*	NL
42144	*	**FD**	A	*GW*	LA	42195		**FG**	P	*GW*	LA
42145	*	**FD**	A	*GW*	LA	42196		**FG**	A	*GW*	LA
42146		**GN**	A	*GN*	EC	42197		**FG**	A	*GW*	LA
42147		**GN**	P	*GN*	EC	42198		**GN**	A	*GN*	EC
42148	u	**MN**	P	*GW*	LA	42199		**GN**	A	*GN*	EC
42149	u	**MN**	P	*GW*	LA	42200	*d	**FD**	A	*GW*	LA
42150		**GN**	A	*GN*	EC	42201	*d	**FD**	A	*GW*	LA
42151	w	**MN**	P	*MM*	NL	42202	d	**FG**	A	*GW*	LA
42152	u	**MN**	P	*MM*	NL	42203		**FG**	A	*GW*	LA
42153	u	**MN**	P	*MM*	NL	42204		**FG**	A	*GW*	LA
42154		**GN**	A	*GN*	EC	42205	u	**MN**	P	*MM*	NL
42155	w	**MN**	P	*MM*	NL	42206	d	**FG**	A	*GW*	LA
42156	u	**MN**	P	*MM*	NL	42207	*d	**FD**	A	*GW*	LA
42157	u	**MN**	P	*MM*	NL	42208		**FG**	A	*GW*	LA
42158	m	**GN**	A	*GN*	EC	42209		**FG**	A	*GW*	LA
42159	m†	**GN**	P	*GN*	EC	42210	u	**MN**	P	*MM*	NL
42160	m	**GN**	P	*GN*	EC	42211	*d	**FD**	A	*GW*	LA
42161	†	**GN**	A	*GN*	EC	42212		**FG**	A	*GW*	LA
42162	*h	**FD**	P	*GW*	LA	42213		**FG**	A	*GW*	LA
42163		**GN**	P	*GN*	EC	42214		**FG**	A	*GW*	LA
42164		**GN**	P	*GN*	EC	42215		**GN**	A	*GN*	EC
42165		**GN**	P	*GN*	EC	42216	*	**FD**	A	*GW*	LA
42166	*h	**FD**	P	*GW*	LA	42217		**FG**	P	*GW*	LA
42167	*h	**FD**	FG	*GW*	LA	42218		**FG**	P	*GW*	LA
42168	*h	**FD**	FG	*GW*	LA	42219		**GN**	A	*GN*	EC
42169	*h	**FD**	FG	*GW*	LA	42220	w	**MN**	P	*MM*	NL
42170	*h	**FD**	P	*GW*	LA	42221	*h	**FD**	P	*GW*	LA
42171		**GN**	A	*GN*	EC	42222	*h	**FD**	P	*GW*	LA
42172		**GN**	A	*GN*	EC	42223	*h	**FD**	P	*GW*	LA
42173		**FG**	P	*GW*	LA	42224		**FG**	P	*GW*	LA
42174		**FG**	P	*GW*	LA	42225	u	**MN**	P	*MM*	NL
42175	*h	**FD**	FG	*GW*	LA	42226		**GN**	A	*GN*	EC
42176	*h	**FD**	FG	*GW*	LA	42227	u	**MN**	P	*MM*	NL
42177	*h	**FD**	FG	*GW*	LA	42228		**GN**	P	*GN*	EC
42178	*h	**FD**	P	*GW*	LA	42229	u	**MN**	P	*MM*	NL
42179		**GN**	A	*GN*	EC	42230	u	**MN**	P	*MM*	NL
42180		**GN**	A	*GN*	EC	42231	*h	**FD**	FG	*GW*	LA
42181		**GN**	A	*GN*	EC	42232	*h	**FD**	FG	*GW*	LA
42182		**GN**	A	*GN*	EC	42233	*h	**FD**	FG	*GW*	LA
42183	*d	**FD**	A	*GW*	LA	42234	s	**V**	P		LM
42184	*	**FD**	A	*GW*	LA	42235		**GN**	A	*GN*	EC
42185	*	**FD**	A	*GW*	LA	42236	*	**FD**	A	*GW*	LA
42186		**GN**	A	*GN*	EC	42237		**GN**	P	*GN*	EC
42187	*d	**FD**	P	*GW*	LA	42238	m†	**GN**	A	*GN*	EC
42188	†	**GN**	A	*GN*	EC	42239	†	**GN**	A	*GN*	EC
42189	m†	**GN**	A	*GN*	EC	42240		**GN**	A	*GN*	EC
42190		**GN**	A	*GN*	EC	42241		**GN**	A	*GN*	EC
42191	m	**GN**	A	*GN*	EC	42242		**GN**	A	*GN*	EC
42192	m	**GN**	A	*GN*	EC	42243		**GN**	A	*GN*	EC

42244		**GN**	A	*GN*	EC	42293	*	**FD**	A	*GW*	LA
42245	*	**FD**	A	*GW*	LA	42294	x	**GN**	P	*GN*	EC
42247	*h	**FD**	P	*GW*	LA	42295	d	**FG**	A	*GW*	LA
42248	*h	**FD**	P	*GW*	LA	42296		**FG**	A	*GW*	LA
42249	*h	**FD**	P	*GW*	LA	42297		**FG**	A	*GW*	LA
42250	*	**FD**	A	*GW*	LA	42299	*d	**FD**	A	*GW*	LA
42251	*k	**FD**	A	*GW*	LA	42300	*	**FD**	A	*GW*	LA
42252	*	**FD**	A	*GW*	LA	42301	*	**FD**	A	*GW*	LA
42253	*	**FD**	A	*GW*	LA	42302		**V**		*FG*	ZI+
42254		**FG**	P	*GW*	LA	42303	*h	**FD**		FG *GW*	LA
42255	d	**FG**	A	*GW*	LA	42304	*h	**FD**		FG *GW*	LA
42256		**FG**	A	*GW*	LA	42305	*h	**FD**		FG *GW*	LA
42257		**FG**	A	*GW*	LA	42306		**GN**	P	*GN*	EC
42258		**FG**	P	*GW*	LA	42307		**GN**	P	*GN*	EC
42259	*k	**FD**	A	*GW*	LA	42308	*h	**FD**	P	*GW*	LA
42260	*	**FD**	A	*GW*	LA	42309	*h	**FD**	P	*GW*	LA
42261	*	**FD**	A	*GW*	LA	42314	*h	**FD**	P	*GW*	LA
42262		**FG**	P	*GW*	LA	42315	*h	**FD**	P	*GW*	LA
42263		**FG**	A	*GW*	LA	42316	*h	**FD**	P	*GW*	LA
42264	*k	**FD**	A	*GW*	LA	42317		**FG**	P	*GW*	LA
42265		**FG**	A	*GW*	LA	42319	*h	**FD**	P	*GW*	LA
42266		**FG**	P	*GW*	LA	42320	*h	**FD**	P	*GW*	LA
42267	*d	**FD**	A	*GW*	LA	42321	*h	**FD**	P	*GW*	LA
42268	*d	**FD**	A	*GW*	LA	42322		**V**	P		ZB
42269	*	**FD**	A	*GW*	LA	42323	m	**GN**	A	*GN*	EC
42271	*d	**FD**	A	*GW*	LA	42324	w	**MN**	P	*MM*	NL
42272	*	**FD**	A	*GW*	LA	42325	*	**FD**	A	*GW*	LA
42273	*	**FD**	A	*GW*	LA	42326		**GN**	P	*GN*	EC
42275	*d	**FD**	A	*GW*	LA	42327	w	**MN**	P	*MM*	NL
42276	*	**FD**	A	*GW*	LA	42328	w	**MN**	P	*GW*	LA
42277	*	**FD**	A	*GW*	LA	42329	w	**MN**	P	*MM*	NL
42279	d	**FG**	A	*GW*	LA	42330		**GN**	P	*GN*	EC
42280		**FG**	A	*GW*	LA	42331	w	**MN**	P	*MM*	NL
42281		**FG**	A	*GW*	LA	42332	*	**FD**	A	*GW*	LA
42283	*	**FD**	A	*GW*	LA	42333	*	**FD**	A	*GW*	LA
42284		**FG**	A	*GW*	LA	42334	*h	**FD**	P	*GW*	LA
42285		**FG**	A	*GW*	LA	42335	u	**MN**	P	*MM*	NL
42286	x	**GN**	P	*GN*	EC	42336	*h	**FD**	P	*GW*	LA
42287	*k	**FD**	A	*GW*	LA	42337	w	**MN**	P	*MM*	NL
42288	*	**FD**	A	*GW*	LA	42338	*h	**FD**	P	*GW*	LA
42289	*	**FD**	A	*GW*	LA	42339	w	**MN**	P	*MM*	NL
42290	s	**V**	P		LM	42340		**GN**	A	*GN*	EC
42291	*d	**FD**	A	*GW*	LA	42341	u	**MN**	P	*GW*	LA
42292	*d	**FD**	A	*GW*	LA						

42342	(44082)	u	**MN**	A	*MM*	NL
42343	(44095)	*	**FD**	A	*GW*	LA
42344	(44092)	*d	**FD**	A	*GW*	LA
42345	(44096)	*d	**FD**	A	*GW*	LA
42346	(41053)	*h	**FD**	A	*GW*	LA
42347	(41054)	*d	**FD**	A	*GW*	LA

42348	(41073)	*k	FD	A	GW	LA
42349	(41074)	*h	FD	A	GW	LA
42350	(41047)		FG	A	GW	LA
42351	(41048)	*	FD	A	GW	LA
42352	(42142, 41176)		GN	P	GN	EC
42353	(42001, 41171)		FT	FG		ZD+
42354	(42114, 41175)		GN	A	GN	EC
42355	(42000, 41172)		GN	A	GN	EC
42356	(42002, 41173)		FG	A	GW	LA
42357	(41002, 41174)		GN	A	GN	EC
42360	(42084, 45084)		FG	A	GW	LA
42361	(44099, 42000)	*	FD	A	GW	LA
42362	(42011, 41178)		FG	A	GW	LA
42363	(41082)	†	GN	A	GN	EC

The following vehicles are converted from loco-hauled Mark 3 vehicles.

42401	(12149)	GC ST	HT		42406	(12112)	GC ST	HT
42402	(12155)	GC ST	HT		42407	(12044)	GC ST	HT
42403	(12033)	GC ST	HT		42408	(12121)	GC ST	HT
42404	(12152)	GC ST	HT		42409	(12088)	GC ST	HT
42405	(12136)	GC ST	HT					

GJ2G (TGS)   TRAILER GUARD'S STANDARD

44000. Lot No. 30953 Derby 1980. 33.47 t.
44001–44090. Lot No. 30949 Derby 1980–82. 33.47 t.
44091–44094. Lot No. 30964 Derby 1982. 33.47 t.
44097–44101. Lot No. 30970 Derby 1982. 33.47 t.

As built and m –/65 1T (w –/63 1T 1W).
*   Refurbished First Great Western vehicles. New Grammer seating and toilet removed for trolley store. –/67 (unless h).
h   "High density" FGW vehicles. –/71.
m   Refurbished GNER "Mallard" vehicles with new Primarius seating.
s   Fitted with centre luggage stack –/63 1T.
t   Fitted with centre luggage stack –/61 1T.

44000	w	FG	P	GW	LA		44016	*h	FD	A	GW	LA
44001	*	FD	A	GW	LA		44017	s	MN	A	MM	NL
44002	*h	FD	A	GW	LA		44018	*	FD	A	GW	LA
44003	*h	FD	A	GW	LA		44019	m	GN	A	GN	EC
44004	w	FG	A	GW	LA		44020	w	FG	A	GW	LA
44005	*	FD	A	GW	LA		44021	t	V	P		LM
44007	w	FG	A	GW	LA		44022	*h	FD	A	GW	LA
44008	w	FG	A	GW	LA		44023	*	FD	A	GW	LA
44009	w	FG	A	GW	LA		44024	w	FG	A	GW	LA
44010	*	FD	A	GW	LA		44025	*	FD	A	GW	LA
44011	w	FG	A	GW	LA		44026	*	FD	A	GW	LA
44012	s	MN	A	MM	NL		44027	s	MN	P	MM	NL
44013	w	FG	A	GW	LA		44028	*	FD	A	GW	LA
44014	w	FG	A	GW	LA		44029	w	FG	A	GW	LA
44015	*	FD	A	GW	LA		44030	*	FD	A	GW	LA

44031	w	**GN**	A	*GN*	EC	44064	w	**FG**	A	*GW*	LA
44032	*	**FD**	A	*GW*	LA	44065	t	**V**	ST		ZJ
44033	*	**FD**	A	*GW*	LA	44066	w	**FG**	A	*GW*	LA
44034	*	**FD**	A	*GW*	LA	44067	w	**FG**	A	*GW*	LA
44035	w	**FG**	A	*GW*	LA	44068	*h	**FD**	FG	*GW*	LA
44036	w	**FG**	A	*GW*	LA	44069	*h	**FD**	P	*GW*	LA
44037	*	**FD**	A	*GW*	LA	44070	s	**MN**	P	*MM*	NL
44038	*	**FD**	A	*GW*	LA	44071	s	**MN**	P	*MM*	NL
44039	w	**FG**	A	*GW*	LA	44072	t	**V**	P		LM
44040	*	**FD**	A	*GW*	LA	44073	s	**MN**	P	*MM*	NL
44041	s	**MN**	P	*MM*	NL	44074	*h	**FD**	FG	*GW*	LA
44042	*h	**FD**	P	*GW*	LA	44075	w	**GN**	P	*GN*	EC
44043	*	**FD**	A	*GW*	LA	44076	*h	**FD**	FG	*GW*	LA
44044	s	**MN**	P	*MM*	NL	44077	w	**GN**	A	*GN*	EC
44045	w	**GN**	A	*GN*	EC	44078	*h	**FD**	P	*GW*	LA
44046	s	**MN**	P	*MM*	NL	44079	*h	**FD**	P	*GW*	LA
44047	s	**MN**	P	*MM*	NL	44080	w	**GN**	A	*GN*	EC
44048	s	**MN**	P	*MM*	NL	44081	*h	**FD**	FG	*GW*	LA
44049	*	**FD**	A	*GW*	LA	44083	w	**GN**	P	*GN*	EC
44050	w	**GN**	P	*GN*	EC	44085	s	**MN**	P	*GW*	LA
44051	s	**MN**	P	*MM*	NL	44086	*	**FD**	A	*GW*	LA
44052	s	**MN**	P	*MM*	NL	44088	t	**V**	ST		ZJ
44054	s	**MN**	P	*MM*	NL	44089	t	**V**	ST		ZJ
44055	*h	**FD**	FG	*GW*	LA	44090	*h	**FD**	P	*GW*	LA
44056	w	**GN**	A	*GN*	EC	44091	*h	**FD**	P	*GW*	LA
44057	m	**GN**	P	*GN*	EC	44093	*h	**FD**	A	*GW*	LA
44058	w	**GN**	A	*GN*	EC	44094	w	**GN**	A	*GN*	EC
44059	w	**FG**	A	*GW*	LA	44097	*	**FD**	P	*GW*	LA
44060	*h	**FD**	P	*GW*	LA	44098	w	**GN**	A	*GN*	EC
44061	m	**GN**	A	*GN*	EC	44100	*h	**FD**	FG	*GW*	LA
44063	w	**GN**	A	*GN*	EC	44101	*h	**FD**	P	*GW*	LA

3. SALOONS

Several specialist passenger carrying vehicles, normally referred to as saloons are permitted to run on the National Rail system. Many of these are to pre-nationalisation designs.

WCJS FIRST SALOON

Built 1892 by LNWR, Wolverton. Originally dining saloon mounted on six-wheel bogies. Rebuilt with new underframe with four-wheel bogies in 1927. Rebuilt 1960 as observation saloon with DMU end. Gangwayed at other end. The interior has a saloon, kitchen, guards vestibule and observation lounge. Gresley bogies. 19/– 1T. 28.5 t. 75 m.p.h.

Non-standard livery: London & North Western Railway.

41   (484, 45018)   x   **0**   SH   *SH*   CJ

LNWR DINING SALOON

Built 1890 by LNWR, Wolverton. Mounted on the underframe of LMS GUV 37908 in the 1980s. Contains kitchen and dining area seating 12 at tables for two. Gresley bogies. 10/–. 75 m.p.h. 25.4 t.

Non-standard livery: London & North Western Railway.

159   (5159)   x   **0**   SH   *SH*   CJ

GNR FIRST CLASS SALOON

Built 1912 by GNR, Doncaster. Contains entrance vestibule, lavatory, two separate saloons, library and luggage space. Gresley bogies. 19/– 1T. 75 m.p.h. 29.4 t.

Non-standard livery: Teak.

807   (4807)   x   **0**   SH   *SH*   CJ

LNER GENERAL MANAGERS SALOON

Built 1945 by LNER, York. Gangwayed at one end with a veranda at the other. The interior has a dining saloon seating 12, kitchen, toilet, office and nine seat lounge. 21/– 1T. B4 bogies. 75 m.p.h. ETH3. 35.7 t.

1999   (902260)   **M**   GS   *GS*   CS   DINING CAR No. 2

GENERAL MANAGER'S SALOON

Renumbered 1989 from London Midland Region departmental series. Formerly the LMR General Manager's saloon. Rebuilt from LMS period 1 BFK M 5033 M to dia. 1654 and mounted on the underframe of BR suburban BS M 43232. Screw couplings have been removed. B4 bogies. 100 m.p.h. ETH2X.

LMS Lot No. 326 Derby 1927. 27.5 t.

6320　(5033, DM 395707) x　**M**　　62　*62*　　SK

GWR FIRST CLASS SALOON

Built 1930 by GWR, Swindon. Contains saloons at either end with body end observation windows, staff compartment, central kitchen and pantry/bar. Numbered DE321011 when in departmental service with British Railways. 20/– 1T. GWR bogies. 75 m.p.h. 34 t.

GWR Lot No. 1431 1930.

9004　　**CH**　RA　*SH*　　CJ

LMS INSPECTION SALOON

Built as engineers' inspection saloons. Non-gangwayed. Observation windows at each end. The interior layout consists of two saloons interspersed by a central lavatory/kitchen/guards section. Now modified for use as an escort coach. BR Mark 1 bogies. 80 m.p.h. 31.5 t.

Lot No. LMS 1356 Wolverton 1944.

45020　　　　**E**　E　*E*　　ML

"QUEEN OF SCOTS" SERVICE CARS

Converted from BR Mark 1 BSKs. Commonwealth bogies. 100 m.p.h. ETH2.

Non-standard livery: London & North Western Railway.

99035. Lot No. 30699 Wolverton 1962–63.
99886. Lot No. 30721 Wolverton 1963.

99035 (35322)　　x　**O**　　SH　*SH*　　CJ　　　SERVICE CAR No. 2
99886 (35407)　　x　**O**　　SH　*SH*　　CJ　　　SERVICE CAR No. 1

VSOE SUPPORT CARS

Converted 1983 (§ 199x) from BR Mark 1 BSK (§ Courier vehicle converted from Mark 1 BSK 1986–87). Toilet retained and former compartment area replaced with train manager's office, crew locker room, linen store and dry goods store. The former luggage area has been adapted for use as an engineers' compartment and workshop. Commonwealth bogies. 100 m.p.h. ETH2.

99538. Lot No. 30229 Metro-Cammell 1955–57. 36 t.
99545. Lot No. 30721 Wolverton 1963. 37 t.

99538 (34991)		**PC**	VS	*VS*	SL	BAGGAGE CAR No. 9
99545 (35466, 80207) §		**PC**	VS	*VS*	SL	BAGGAGE CAR No. 11

VSOE BRAKE LUGGAGE VAN

Converted 199x from BR Mark 1 BG. Guard's compartment retained and former baggage area adapted for stowage of passengers' luggage. pg. B4 bogies. 100 m.p.h. ETH1X.

Lot No. 30162 Pressed Steel 1956–57. 30.5 t.

99554 (80867, 92904)	**VN**	VS		CP

ROYAL SCOTSMAN SALOONS

Built 1960 by Metro-Cammell as Pullman Parlour First (§ Pullman Kitchen First) for East Coast Main Line services. Rebuilt 1990 as sleeping cars with four twin sleeping rooms (*§ three twin sleeping rooms and two single sleeping rooms at each end). Commonwealth bogies. 38.5 t.

99961 (324 AMBER) *	**M**	GS	*GS*	CS	STATE CAR 1
99962 (329 PEARL)	**M**	GS	*GS*	CS	STATE CAR 2
99963 (331 TOPAZ)	**M**	GS	*GS*	CS	STATE CAR 3
99964 (313 FINCH) §	**M**	GS	*GS*	CS	STATE CAR 4

Built 1960 by Metro-Cammell as Pullman Kitchen First for East Coast Main Line services. Rebuilt 1990 as observation car with open verandah seating 32. Commonwealth bogies. 38.5 t.

99965 (319 SNIPE)	**M**	GS	*GS*	CS	OBSERVATION CAR

Built 1960 by Metro-Cammell as Pullman Kitchen First for East Coast Main Line services. Rebuilt 1993 as dining car. Commonwealth bogies. 38.5 t.

99967 (317 RAVEN)	**M**	GS	*GS*	CS	DINING CAR

Mark 3A. Converted from SLEP at Carnforth Railway Restoration and Engineering Services in 1997. BT10 bogies. Attendant's and adjacent two sleeping compartments converted to generator room containing a 160 kW Volvo unit. In 99968 four sleeping compartments remain for staff use with another converted for use as a staff shower and toilet. The remaining five sleeping compartments have been replaced by two passenger cabins. In 99969 seven sleeping compartments remain for staff use. A further sleeping compartment, along with one toilet, have been converted to store rooms. The other two sleeping compartments have been combined to form a crew mess. ETH7X. 41.5 t.

Lot. No. 30960 Derby 1981–3.

99968 (10541)	**M**	GS	*GS*	CS	STATE CAR 5
99969 (10556)	**M**	GS	*GS*	CS	SERVICE CAR

RAILFILMS "LMS CLUB CAR"

Converted from BR Mark 1 TSO at Carnforth Railway Restoration and Engineering Services in 1994. Contains kitchen, pantry and two dining saloons. 20/– 1T. Commonwealth bogies. 100 m.p.h. ETH 4.

Lot. No. 30724 York 1963. 37 t.

99993 (5067)   **M**   RA   *WC*   CS   LMS CLUB CAR

BR INSPECTION SALOON

Mark 1. Short frames. Non-gangwayed. Observation windows at each end. The interior layout consists of two saloons interspersed by a central lavatory/ kitchen/guards/luggage section. Now modified for use as an escort coach. BR Mark 1 bogies. 90 m.p.h.

Lot No. BR Wagon Lot. 3379 Swindon 1960. 30.5 t.

999509   **E**   E   *E*   ML

4. PULLMAN CAR COMPANY SERIES

Pullman cars have never generally been numbered as such, although many have carried numbers, instead they have carried titles. However, a scheme of schedule numbers exists which generally lists cars in chronological order. In this section those numbers are shown followed by the car's title. Cars described as "kitchen" contain a kitchen in addition to passenger accommodation and have gas cooking unless otherwise stated. Cars described as "parlour" consist entirely of passenger accommodation. Cars described as "brake" contain a compartment for the use of the guard and a luggage compartment in addition to passenger accommodation.

PULLMAN PARLOUR FIRST

Built 1927 by Midland Carriage & Wagon Company. Gresley bogies. 26/– 2T. ETH 2. 41 t.

213 MINERVA   **PC** VS *VS* SL

PULLMAN PARLOUR FIRST

Built 1928 by Metropolitan Carriage & Wagon Company. Gresley bogies. 24/– 2T. ETH 4. 40 t.

239 AGATHA   **PC** VS   SL
243 LUCILLE   **PC** VS *VS* SL

PULLMAN KITCHEN FIRST

Built 1925 by BRCW. Rebuilt by Midland Carriage & Wagon Company in 1928. Gresley bogies. 20/– 1T. ETH 4. 41 t.

245 IBIS   **PC** VS *VS* SL

PULLMAN PARLOUR FIRST

Built 1928 by Metropolitan Carriage & Wagon Company. Gresley bogies. 24/– 2T. ETH 4.

254 ZENA   **PC** VS *VS* SL

PULLMAN KITCHEN FIRST

Built 1928 by Metropolitan Carriage & Wagon Company. Gresley bogies. 20/– 1T. ETH 4. 42 t.

255 IONE   **PC** VS *VS* SL

PULLMAN KITCHEN COMPOSITE

Built 1932 by Metropolitan Carriage & Wagon Company. Originally included in 6-Pul EMU. Electric cooking. EMU bogies. 12/16 1T.

264	RUTH	**PC**	VS		SL

PULLMAN KITCHEN FIRST

Built 1932 by Metropolitan Carriage & Wagon Company. Originally included in "Brighton Belle" EMUs but now used as hauled stock. Electric cooking. B5 (SR) bogies (§ EMU bogies). 20/– 1T. ETH 2. 44 t.

280	AUDREY		**PC**	VS	*VS*	SL
281	GWEN		**PC**	VS	*VS*	SL
283	MONA	§	**PC**	VS		SL
284	VERA		**PC**	VS	*VS*	SL

PULLMAN PARLOUR THIRD

Built 1932 by Metropolitan Carriage & Wagon Company. Originally included in "Brighton Belle" EMUs. EMU bogies. –/56 2T.

285	CAR No. 85	**PC**	VS	SL
286	CAR No. 86	**PC**	VS	SL

PULLMAN BRAKE THIRD

Built 1932 by Metropolitan Carriage & Wagon Company. Originally driving motor cars in "Brighton Belle" EMUs. Traction and control equipment removed for use as hauled stock. EMU bogies. –/48 1T.

288	CAR No. 88	**PC**	VS	SL
292	CAR No. 92	**PC**	VS	SL
293	CAR No. 93	**PC**	VS	SL

PULLMAN PARLOUR FIRST

Built 1951 by Birmingham Railway Carriage & Wagon Company. Gresley bogies. 32/– 2T. ETH 3. 39 t.

301	PERSEUS	**PC**	VS	*VS*	SL

Built 1952 by Pullman Car Company, Preston Park using underframe and bogies from 176 RAINBOW, the body of which had been destroyed by fire. Gresley bogies. 26/– 2T. ETH 4. 38 t.

302	PHOENIX	**PC**	VS	*VS*	SL

PULLMAN PARLOUR FIRST

Built 1951 by Birmingham Railway Carriage & Wagon Company. Gresley bogies. 32/– 2T. ETH 3. 39 t.

308 CYGNUS   **PC**   VS   *VS*   SL

PULLMAN FIRST BAR

Built 1951 by Birmingham Railway Carriage & Wagon Company. Rebuilt 1999 by Blake Fabrications, Edinburgh with original timber-framed body replaced by a new fabricated steel body. Contains kitchen, bar, dining saloon and coupé. Electric cooking. Gresley bogies. 14/– 1T. ETH 3.

310 PEGASUS   **PC**   RA   *WT*   OM

Also carries "THE TRIANON BAR" branding.

PULLMAN KITCHEN SECOND

Built 1960–1961 by Metro-Cammell for East Coast Main Line services. Commonwealth bogies. –/30 1T. 40 t.

335 CAR No. 335   x **PC**   VT   *VT*   TM

PULLMAN PARLOUR SECOND

Built 1960–1961 by Metro-Cammell for East Coast Main Line services. Commonwealth bogies. –/42 2T. 38.5 t.

348 CAR No. 348   x **M**   WC   *WC*   CS
349 CAR No. 349   x **PC**   VT   *VT*   TM
350 CAR No. 350   x **M**   WC   *WC*   CS
353 CAR No. 353   x **PC**   VT   *VT*   TM

PULLMAN SECOND BAR

Built 1960–1961 by Metro-Cammell for East Coast Main Line services. Commonwealth bogies. –/24 + 17 bar seats. 38.5 t.

354 THE HADRIAN BAR x **PC**   WC   *WC*   CS

5. LOCOMOTIVE SUPPORT COACHES

These carriages have been adapted from Mark 1 BCK, BFK, BSK, NNX and Mark 2 BFK for use as support carriages for heritage steam locomotives. Some seating is retained for the use of personnel supporting the locomotives operation with the remainder of the carriage adapted for storage, workshop, dormitory and catering purposes. These carriages can spend considerable periods of time off the national railway system when the locomotives they support are not being used on that system. After the depot code, the locomotive(s) each carriage is usually used to support is given. Operator codes are not included in this section. Seating capacities refer to the original vehicle as running in normal service.

AB11 (BFK)   CORRIDOR BRAKE FIRST

Mark 1. 24/– 1T. Commonwealth bogies. ETH 2.

14007. Lot No. 30382 Swindon 1959. 35 t.
17013–17019. Lot No. 30668 Swindon 1961. 36 t.

14007	(14007, 17007)	x **M**	B1	BH	LNER 61264
17013	(14013)	**PC**	JH	SO	LNER 60019
17019	(14019)	x **M**	92	TM	SR 30777

AB1Z (BFK)   CORRIDOR BRAKE FIRST

Mark 2. Pressure ventilated. 24/– 1T. B4 bogies. ETH 4.

Lot No. 30756 Derby 1966. 31.5 t.

17041	(14041)	**M**	DG	BQ	BR 71000

AB1A (BFK)   CORRIDOR BRAKE FIRST

Mark 2A. Pressure ventilated. 24/– 1T. B4 bogies. ETH 4.

Lot No. 30786 Derby 1968. 32 t.

17096	(14096)	**G**	MN	SL	SR 35028

AB31 (BCK)   CORRIDOR BRAKE COMPOSITE

Mark 1. Two first class and three standard class compartments. 12/18 2T (* 12/24 2T). ETH 2.

21232. Lot No. 30574 GRCW 1960. B4 bogies. 34 t.
21268. Lot No. 30732 Derby 1964. Commonwealth bogies. 37 t.

21232	x **CC**	62	SK	LMS 46233
21268	*	BS	SO	LMS 46100

AB21 (BSK)   CORRIDOR BRAKE STANDARD

Mark 1. Four compartments. Metal window frames and melamine interior panelling. –/24 1T. ETH2.

35317–35333. Lot No. 30699 Wolverton 1962–63. Commonwealth bogies. 37 t.
35449. Lot No. 30728 Wolverton 1963. Commonwealth bogies. 37 t.
35453–35486. Lot No. 30721 Wolverton 1963. Commonwealth bogies. 37 t.

35317	x	**G**	IR	BQ	SR 30850
35329	v	**M**	MH	RL	Mid Hants Railway-based locomotives
35333	x	**CH**	24	DI	GWR 6024
35449	x	**M**	BE	BQ	LMS 45231
35453	x	**CH**	GW	DI	GWR 5051
35461	x	**CH**	RV	OM	GWR 5029
35463	v	**M**	WC	CS	GWR 5972/LMS 48151
35465	x	**CC**	LW	CP	LMS 46201
35468	v	**M**	NM	YK	National Railway Museum locomotives
35470	v	**CH**	VT	TM	Tyseley Locomotive Works-based locomotives
35476	x	**M**	62	SK	LMS 46233
35479	v	**M**	SV	KR	Severn Valley Railway-based locomotives
35486	x	**M**	JC	TN	LNER 60009/61264

AB1C (BFK)   CORRIDOR BRAKE FIRST

Mark 2C. Pressure ventilated. Renumbered when declassified. –/24 1T. B4 bogies. ETH 4.

Lot No. 30796 Derby 1969–70. 32.5 t.

35508	(14128, 17128)	**M**	IR	BQ	LMS 45407/BR 76079

AB1A (BFK)   CORRIDOR BRAKE FIRST

Mark 2A. Pressure ventilated. Renumbered when declassified. –/24 1T. B4 bogies. Cage removed from brake compartment. ETH 4.

Lot No. 30786 Derby 1968. 32 t.

35517	(14088, 17088)	**M**	IR	BQ	LMS 45407/BR 76079
35518	(14097, 17097)	**PC**	WC	SO	SR 34067

NNX   COURIER VEHICLE

Mark 1. Converted 1986–7 from BSKs. One compartment and toilet retained for courier use. One set of roller shutter doors inserted on each side. ETH 2.

80204/17. Lot No. 30699 Wolverton 1962. Commonwealth bogies. 37 t.
80220. Lot No. 30573 Gloucester 1960. B4 bogies. 33 t.

80204	(35297)	**M**	WC	CS	GWR 5972/LMS 48151
80217	(35299)	**M**	WC	CS	GWR 5972/LMS 48151
80220	(35276)	**M**	NE	NY	LNER 62005

6. 99xxx RANGE NUMBER CONVERSION TABLE

The following table is presented to help readers identify vehicles which may carry numbers in the 99xxx range, the former private owner number series which is no longer in general use.

99xxx	BR No.	99xxx	BR No.	99xxx	BR No.
99040	21232	99348	Pullman 348	99671	548
99041	35476	99349	Pullman 349	99672	549
99052	Saloon 41	99350	Pullman 350	99673	550
99121	3105	99353	Pullman 353	99674	551
99125	3113	99354	Pullman 354	99675	552
99127	3117	99361	Pullman 335	99676	553
99128	3130	99371	3128	99677	586
99131	1999	99530	Pullman 301	99678	504
99141	17041	99531	Pullman 302	99679	506
99241	35449	99532	Pullman 308	99680	17102
99304	21256	99534	Pullman 245	99710	18767
99311	1882	99535	Pullman 213	99716	18808
99312	35463	99536	Pullman 254	99718	18862
99317	3766	99537	Pullman 280	99721	18756
99318	4912	99539	Pullman 255	99722	18806
99319	17168	99541	Pullman 243	99723	35459
99321	5299	99543	Pullman 284	99792	17019
99326	4954	99546	Pullman 281	99880	159
99327	5044	99547	Pullman 292	99881	807
99328	5033	99548	Pullman 293	99953	35468
99329	4931	99670	546		

PLATFORM 5 MAIL ORDER

FREIGHTMASTER
Freightmaster Publishing

Freightmaster is the Great Britain National Railfreight Timetable. It contains full timetable listings for over 70 key locations around the country, including dates of operation, train type and booked motive power for every train. Most locations feature 0700-2300 listings, with full 24 hour timetables for busy locations. Also includes a separate analysis of national freight flows. Well illustrated by a series of detailed maps.160 pages. **£12.95**

Note: Freightmaster is published 4 times a year in January, April, July and October. Customers ordering this title will be supplied with the latest edition available unless requested otherwise.

LINE BY LINE
Freightmaster Publishing

Line by Line is a series of excellent guidebooks tracing the route of Britain's main line railways. Each page covers a five-mile section of route with gradient profiles, track layout diagrams and a black & white illustration provided for each section. Also includes a general overview of the line, a gallery section of colour photographs, several OS map reproductions and a table of distances in miles & chains. The following volumes are currently available:

Line by Line: The Scottish Highland Lines **£17.95**
Line by Line: The East Coast Main Line **£14.95**
Line by Line: The Midland Route **£14.95**

Please add postage: 10% UK, 20% Europe, 30% Rest of World.

Telephone, fax or send your order to the Platform 5 Mail Order Department. See inside back cover of this book for details.

Overhead Line Equipment Test Coach ("MENTOR"). Can either be locomotive-hauled or included between DMU vehicles 977391/2. Converted from BR Mark 1 BSK Lot No. 30142 Gloucester 1954–5. Fitted with pantograph. B4 bogies.

975091 (34615)	**Y**	NR	*SO*	ZA

Structure Gauging Train Dormitory and Generator Coach. Converted from BR Mark 1 BCK Lot No. 30732 Derby 1962–4. B4 bogies.

975280 (21263)	**Y**	NR	*SO*	ZA

Test Coach. Converted from BR Mark 2 FK Lot No. 30734 Derby 1962–64. B4 bogies.

975290 (13396)	**SO**	SO		ZA

Test Coach. Converted from BR Mark 1 BSK Lot No. 30699 Wolverton 1961–63. Commonwealth bogies.

975397 (35386)	**SO**	SO	*SO*	ZA

New Measurement Train Conference Coach. Converted from prototype HST TF Lot No. 30848 Derby 1972. BT10 bogies.

975814 (11000,41000)	**Y**	NR	*SO*	EC

New Measurement Train Lecture Coach. Converted from prototype HST TRUB Lot No. 30849 Derby 1972–3. BT10 bogies.

975984 (10000, 40000)	**Y**	NR	*SO*	EC

Track Recording Train Dormitory Coach. Converted from BR Mark 2 BSO. Lot No 30757 Derby 1965–66. B4 bogies.

977337 (9395)	**Y**	NR	*SO*	ZA

Test Train Brake & Stores Coach. Converted from Mark 2 BSO. Lot No. 30757 Derby 1965–66. B4 bogies.

977338 (9387)	**SO**	SO		ZA

Radio Equipment Survey Coaches. Converted from BR Mark 2E TSO. Lot No. 30844 Derby 1972–73. B4 bogies.

977868 (5846)	**Y**	NR	*SO*	ZA
977869 (5858)	**Y**	NR	*SO*	ZA

Test Train Staff Coach. Converted from Royal Household couchette Lot No. 30889, which in turn had been converted from BR Mark 2B BFK Lot No. 30790 Derby 1969. B5 bogies.

977969 (14112, 2906)	**Y**	NR	*SO*	ZA

Laboratory Coach. Converted from BR Mark 2E TSO. Lot No. 30844 Derby 1972–73. B4 bogies.

977974 (5854)	**Y**	DE		ZA

Hot Box Detection Coach. Converted from BR Mark 2F FO converted to Class 488/2 EMU TFOH. Lot No. 30859 Derby 1973–74. B4 bogies.

977983 (3407, 72503)	**RK**	NR	*SO*	ZA

New Measurement Train Staff Coach. Converted from HST TRFK. Lot No. 30884 Derby 1976–77. BT10 bogies.

977984 (40501)   **Y**   P   *SO*   EC

Structure Gauging Train Coach. Converted from BR Mark 2F TSO converted to Class 488/3 EMU TSO. Lot No. 30860 Derby 1973–74. B4 bogies.

977985 (6019, 72715)   **Y**   NR   *SO*   ZA

Track Recording Train Coach. Converted from BR Mark 2D FO subsequently declassified to SO and then converted to exhibition van. Lot No. 30821 Derby 1971.

977986 (3189, 99664)   **Y**   NR   *SO*   ZA

New Measurement Train Overhead Line Equipment Test Coach. Converted from HST TGS. Lot No. 30949 Derby 1982. Fitted with pantograph. BT10 bogies.

977993 (44053)   **Y**   P   *SO*   EC

New Measurement Train Track Recording Coach. Converted from HST TGS. Lot No. 30949 Derby 1982. BT10 bogies.

977994 (44087)   **Y**   P   *SO*   EC

New Measurement Train Coach. Converted from HST TRFM. Lot No. 30921 Derby 1978–79. BT10 bogies. Fitted with generator.

977995 (40719, 40619)   **Y**   P   *SO*   EC

New Measurement Train Coach. Converted from HST TGS for Hitachi. Lot No. 30949 Derby 1982. Fitted with batteries for use with HST power car 43089 as part of the New Measurement Train. BT10 bogies.

977996 (44062)   **Y**   HI   *SO*   EC

Inspection Coach. Converted from BR Inspection Saloon. BR Wagon Lot No. 3095. Swindon 1957. B4 bogies.

999506 AMANDA   **M**   NR   *SO*   ZA

Track Recording Coach. Converted from BR Inspection Saloon. BR Wagon Lot No. 3379. Swindon 1960. B4 bogies.

999508   **Y**   SO   *SO*   ZA

Track Recording Coach. Purpose built Mark 2. B4 bogies.

999550   **Y**   NR   *SO*   ZA

Ultrasonic Test Coach. Converted from a Class 421 EMU MBSO. Lot No. 30816. York 1970. ? bogies.

999606 (62356)   **Y**   NR   *SO*   ZA

▲ First Great Western Green-liveried Mark 3A Sleeping Car 10594 is seen at Penzance on 04/07/07. **Robert Pritchard**

▼ Royal Train-liveried Mark 3B Royal Saloon 2923 is seen at Hexthorpe, Doncaster on 13/07/06. **Robert Pritchard**

▲ Carrying the new First Great Western Dynamic Lines livery, HST TS 42231 is seen near Teignmouth on 01/09/07. **Robert Pritchard**

▼ Refurbished GNER-liveried HST TS 42292 stands at Paignton on 01/09/07 in the formation of a Virgin Cross-Country service to Newcastle. **Robert Pritchard**

▲ Midland Mainline-liveried HST TGS 44054 is seen at Sheffield on 31/08/07.
Robert Pritchard

▼ GNER-liveried Mark 4 TSO (end) 12227 is seen at Doncaster on 09/09/07.
Robert Pritchard

▲ Royal Scotsman Saloon 99963 "STATE CAR No. 3" is seen passing Attadale on 21/08/07.   **Robert Pritchard**

▼ Pullman Kitchen First 245 "IBIS" is seen at Restormel on 06/07/07, forming part of the VSOE British Pullman train which is based at Stewarts Lane, London.
                                                                                Robert Pritchard

▲ "One"-liveried Mark 3B DVT 82107 is seen at Colchester forming the 09.30 Liverpool Street–Norwich on 11/08/07 (90009 was providing power at the rear). **Andrew Mist**

▼ GNER-liveried Mark 4 DVT 82225 is seen leading the 12.00 Glasgow Central–London King's Cross through Welwyn North on 01/06/07. **Robert Pritchard**

▲ BR Southern Region Green-liveried Network Rail Inspection Saloon 975025 "CAROLINE" is seen being propelled by 33103 through Knockholt en route to Hastings via Ashford, Maidstone East and Victoria on 09/09/05. **Rodney Lissenden**

▼ BR Maroon-liveried Inspection Coach 999506 "CAROLINE" is seen near Cardiff Central being hauled by 47854 as a 2Z01 09.30 Cheltenham–Llanelli on 17/04/07. **Andrew Mist**

▲ Network Rail yellow-liveried Laboratory Coach 977974 (converted from Mark 2E 5854) is seen at Quorn & Woodhouse (Great Central Railway) on 03/05/07.
Robert Pritchard

▼ Railtrack blue-liveried Hot Box Detection Coach 977983 is seen at Cardiff Central on 15/09/06, as part of a Swansea–Derby test train. **Andrew Mist**

▲ Network Rail yellow-liveried 977996 was converted from HST TGS 44062 at Brush, Loughborough in 2007 to act as a battery coach to run with 43089 as part of a Hitachi "hybrid" HST test set. It is seen on test at Quorn & Woodhouse on the Great Central Railway on 03/05/07. **Robert Pritchard**

▼ Internal User vehicle 041947 (former GUV 93425) is seen at Ilford depot on 27/09/06. **Robert Pritchard**

TEST TRAIN BRAKE FORCE RUNNER SETS

Converted from Class 488/3 ex-Gatwick Express locomotive hauled stock (formerly Mark 2 coaches). These sets of vehicles are included in test trains to provide brake force and are not used for any other purposes.

72612–72616/72630/72631/72639. Lot No. 30860 Derby 1973–74. 33.5 t.
72708. Lot No. 30860 Derby 1973–74. 33.5 t.

910 001	**RK**	NR	*SO*	ZA	72616 (6007)	72708 (6095)	72639 (6070)
910 002	**RK**	NR	*SO*	ZA	72612 (6156)		72613 (6126)
-	**Y**	NR	*SO*	ZA	72630 (6094)		72631 (6096)

BREAKDOWN TRAIN COACHES

These coaches are formed in trains used for the recovery of derailed railway vehicles and were converted from BR Mark 1 BCK, BG, BSK and SK. The current use of each vehicle is given. 975611–613 were previously converted to trailer luggage vans in 1968. BR Mark 1 bogies.

975080. Lot No. 30155 Wolverton 1955–56.
975087. Lot No. 30032 Wolverton 1951–52.
975463/573. Lot No. 30156 Wolverton 1954–55.
975465/477/494. Lot No. 30233 GRCW 1955–57.
975471. Lot No. 30095 Wolverton 1953–55.
975481/482/574. Lot No. 30141 GRCW 1954–55.
975498. Lot No. 30074 Wolverton 1953–54.
975611–613. Lot No. 30162 Pressed Steel 1954–57.
977088/235. Lot No. 30229 Metro-Cammell 1955–57.
977107. Lot No. 30425 Metro-Cammell 1956–58.

r refurbished

975080	(25079)	r	**Y**	NR	*E*	OM	Tool Van
975087	(34289)	r	**NR**	NR	*E*	OM	Generator Van
975463	(34721)	r	**Y**	NR	*E*	TE	Staff Coach
975465	(35109)	r	**Y**	NR	*E*	OM	Staff Coach
975471	(34543)	r	**NR**	NR	*E*	OM	Staff & Tool Coach
975477	(35108)	r	**NR**	NR	*E*	OM	Staff Coach
975481	(34606)	r	**Y**	NR	*E*	OM	Generator Van
975482	(34602)	r	**Y**	NR	*E*	TE	Generator Van
975494	(35082)	r	**Y**	NR	*E*	MG	Generator Van
975498	(34367)	r	**Y**	NR	*E*	TE	Tool Van
975573	(34729)	r	**Y**	NR	*E*	MG	Staff Coach
975574	(34599)	r	**Y**	NR	*E*	OM	Staff Coach
975611	(80915, 68201)	r	**Y**	NR	*E*	OM	Generator Van
975612	(80922, 68203)	r	**Y**	NR	*E*	MG	Tool Van
975613	(80918, 68202)	r	**Y**	NR	*E*	OM	Tool Van
977088	(34990)		**Y**	NR	*E*	CE	Generator Van
977107	(21202)		**Y**	NR	*E*	CE	Staff Coach
977235	(34989, 083172)		**Y**	NR	*E*	CE	Tool Van

INFRASTRUCTURE MAINTENANCE COACHES

Overhead Line Maintenance Coaches

These coaches were formed in a train used for the maintenance, repair and renewal of overhead lines and were converted from BR Mark 1 BSK, CK and SK. The former use of each vehicle is given. All have been refurbished.

Non-standard livery: Light grey with a blue stripe.

975699. Lot No. 30233 GRCW 1955–57. BR Mark 1 bogies.
975700. Lot No. 30025 Wolverton 1950–52. BR Mark 1 bogies.
975714. Lot No. 30374. York 1958. Commonwealth bogies.
975724. Lot No. 30471 Metro-Cammell 1957–59. Commonwealth bogies.
975734. Lot No. 30426 Wolverton 1956–58. BR Mark 1 Bogies.
975744. Lot No. 30350 Wolverton 1956–57.BR Mark 1 bogies.

975699	(35105)	**0**	NR	Preston	Pantograph coach
975700	(34138)	**0**	NR	Preston	Pantograph coach
975714	(25466)	**0**	NR	Preston	Stores van
975724	(16079)	**0**	NR	Preston	Stores & generator van
975734	(25695)	**0**	NR	Preston	Stores & roof access coach
975744	(25440)	**0**	NR	Preston	Staff & office coach

Snowblower Train Coaches

These coaches worked with Snowblower ADB 968501. They were converted from BR Mark 1 BSK. The former use of each vehicle is given. Commonwealth bogies.

975464. Lot No. 30386 Charles Roberts 1956–58.
975486. Lot No. 30025 Wolverton 1950–52.

975464	(35171)	**Y**	NR	ZR	Staff & dormitory coach
975486	(34100)	**Y**	NR	ZR	Tool van

Snowblower Train Tool Vans

These vans worked with Snowblower ADB 968500.

200715. Wagon Lot No. 3855 Ashford 1976. 4-wheeled.
787395. Wagon Lot No. 3567 Eastleigh 1966. 4-wheeled.

200715	**Y**	NR	IS
787395	**Y**	NR	IS

De-Icing Coaches

These coaches are used for removing ice from the conductor rail of DC lines. They were converted from Class 489 DMLVs that had originally been Class 414/3 DMBSOs.

Lot No. 30452 Ashford/Eastleigh 1959. Mk 4 bogies.

68501	(61281)	**Y**	NR	*GB*	TW
68508	(61272)	**Y**	NR	*GB*	TW

Miscellaneous Infrastructure Coaches

These coaches are used for various infrastructure projects on the National Railway network. They were converted from BR Mark 1 BSK & BG and BR Mark 3 SLEP. The current use of each vehicle is given.

977163/165/166. Lot No. 30721 Wolverton 1961–63. Commonwealth bogies.
977167. Lot No. 30699 Wolverton 1961–63. Commonwealth bogies.
977168. Lot No. 30573 GRCW 1959–60. B4 bogies.
977989. Lot No. 30960 Derby 1981–83. BT 10 bogies.
977990. Lot No. 30228 Metro-Cammell 1957-58. B4 bogies.
977991. Lot No. 30323 Pressed steel 1957. B4 bogies.

Non-standard liveries:

977163 and 977167 White with a blue stripe.
977165, 975166 and 975168 All over white.

977163	(35487)	**O**	BB	*BB*	AP	Staff & generator coach
977165	(35408)	**O**	BB	*BB*	AP	Staff & generator coach
977166	(35419)	**O**	BB	*BB*	AP	Staff & generator coach
977167	(35400)	**O**	BB	*BB*	AP	Staff & generator coach
977168	(35289)	**O**	BB	*BB*	AP	Staff & generator coach
977989	(10536)	**M**	J		Leeman Road EY, York	Staff & Dormitory Coach
977990	(81165, 92937)	**NR**	NR		OM	Tool Van
977991	(81308, 92991)	**NR**	NR		OM	Tool Van

INTERNAL USER VEHICLES

These vehicles are confined to yards and depots or do not normally move at all. Details are given of the internal user number (if allocated), type and former identity, current use and location. Many of those listed no longer see regular use.

024877	BR CCT 94698	Stores van	Wavertree Yard, Edge Hill
024909	BR BSOT 9106	Staff accommodation	Preston Station
025000	BR BSO 9423	Staff accommodation	Preston Station
025026	BR TSO 5259	Staff accommodation	Wavertree Yard, Edge Hill
041379	LMS CCT 35527	Stores van	Leeman Road EY, York
041898	BR BG 84608	Stores van	Leeman Road EY, York
041947	BR GUV 93425	Stores van	IL
041963	LMS milk tank 44047	Storage tank	DR
042154	BR GUV 93975	Stores van	Ipswich Upper Yard
061061	BR CCT 94135	Stores van	Oxford station
061223	BR GUV 93714	Stores van	Oxford station
083439	BR CCT 94752	Stores van	WD
083602	BR CCT 94494	Stores van	Three Bridges station
083633	BR GUV 93724	Stores van	BI
083637	BR NW 99203	Stores van	SL
083644	BR Ferry Van 889201	Stores van	EH
083664	BR Ferry Van 889203	Stores van	EH
095020	LNER BG 70170	Stores van	Inverness Yard
095030	BR GUV 96140	Stores van	EC
-	BR FO 3186	Instruction Coach	DY
-	BR FO 3381	Instruction Coach	HE
-	BR TSO 5636	Instruction Coach	PM
-	BR HSBV 6396	Stores van	MA
-	BR BFK 17156	Instruction Coach	DY
-	BR TSOLH 72614	Instruction Coach	DY
-	BR TSOLH 72615	Instruction Coach	DY
-	BR TSOL 72707	Instruction Coach	DY
-	BR BG 92901	Stores van	WB
-	BR GUV 93722	Stores van	SL
-	BR NL 94003	Stores van	OO
-	BR NL 94006	Stores van	OO
-	BR NB 94101	Stores van	GL
-	BR NK 94121	Stores van	TO
-	BR CCT 94181	Stores van	SL
-	BR GUV 96139	Stores van	MA
-	BR Ferry Van 889017	Stores van	SL
-	BR Ferry Van 889200	Stores van	SL
-	BR Ferry Van 889202	Stores van	SL
-	BR TSO 975403	Cinema Coach	PM

Notes: CCT = Covered Carriage Truck (a 4-wheeled van similar to a GUV).

NL = Newspaper Van (converted from a GUV).

9. COACHING STOCK AWAITING DISPOSAL

This list contains the last known locations of coaching stock awaiting disposal. The definition of which vehicles are "awaiting disposal" is somewhat vague, but generally speaking these are vehicles of types not now in normal service or vehicles which have been damaged by fire, vandalism or collision.

1644	CS	6145	CT	10709	ZN	92146	ZA
1650	CS	6157	CT	11024	ZH	92159	KT
1652	CS	6161	CT	11037	ZD	92174	SN
1655	CS	6167	CT	12023	ZD	92175	CP
1663	CS	6178	HM	12070	ZN	92193	**
1670	CS	6335	LA	12096	ZN	92194	SN
1688	CS	6339	EC	13306	CS	92198	CT
2127	CS	6345	EC	13320	CS	92303	OM
3181	CD	6356	CV	13321	CS	92350	OM
3429	AS	6357	CV	13323	CS	92384	OM
3521	BR	6360	NL	13508	CS	92400	CD
4849	CD	6361	NL	17144	CV	92530	OM
4854	CD	6378	LM	17153	CS	92908	CS
4860	CS	6379	LM	17161	OM	92929	CD
4932	CS	6523	CS	17169	CV	92931	SN
4997	CS	6800	CT	17170	CV	92935	SN
5267	KT	6804	CT	18837	CS	92936	CD
5354	SN	6805	ZG	18893	CS	92938	SN
5389	OM	6806	LM	19013	CS	92939	ZA
5420	OM	6808	ZG	34525	CS	93180	**
5446	KT	6811	ZG	35513	CD	93723	BY
5480	KT	6817	ZG	35516	CD	94103	SD
5505	CS	6820	LM	45029	ML	94104	OM
5600	CS	6821	ZG	68504	BH	94106	ML
5647	CD	6823	LM	68505	BH	94113	OM
5704	CS	6824	KT	80211	DY	94116	TY
5714	CS	6828	ZG	80349	WE	94137	ML
5727	CS	6829	ZG	80401	TO	94147	ML
5756	CS	6900	**	80403	CS	94150	BS
5773	CT	6901	**	80404	CS	94153	WE
5781	NC	9482	NL	80432	TO	94155	BS
5824	CT	9712	ZJ	80434	TO	94160	ML
5827	CT	10201	ZB	80438	WE	94166	BS
5845	CT	10327	ZC	80458	OM	94168	BS
5852	CT	10515	IS	81025	CP	94170	ML
5918	CT	10540	OM	84364	DW	94176	BS
5951	CT	10547	IS	84519	OM	94177	SM
6047	CT	10554	TO	92100	CP	94190	BK
6100	CT	10663	IS	92111	CD	94191	ML
6144	CT	10682	TO	92114	DY	94192	BS

94195	BS	94413	ML	94532	OM	96605	LM
94196	ML	94416	TY	94534	ML	96606	LM
94197	BS	94420	ML	94536	ML	96607	LM
94199	ML	94422	OM	94538	ML	96608	LM
94203	ML	94423	BS	94539	ML	96609	LM
94207	OM	94427	WE	94540	TJ	99014	**
94208	SM	94428	BS	94541	ML	99015	**
94209	SD	94429	TE	94542	TY	99019	ZR
94213	ML	94431	ML	94543	BS	99025	KT
94214	ML	94432	TY	94544	ML	99026	KT
94217	ML	94433	AC	94545	TE	99027	KT
94221	ML	94434	TY	94546	TY	99646	SL
94222	ML	94435	OM	94547	ML	889026	**
94224	CD	94438	TO	94548	TY	975000	ZA
94225	ML	94440	TY	95228	NC	975379	**
94227	TE	94445	WE	95300	ML	975454	TO
94229	ML	94450	WE	95301	ML	975484	CS
94302	TY	94451	WE	95400	ML	975497	**
94303	TY	94458	SD	95410	ML	975535	**
94304	ML	94462	CD	95727	WE	975615	SJ
94306	TY	94463	TY	95754	TY	975639	CS
94307	SD	94470	OM	95758	ML	975681	**
94308	ML	94476	CD	95761	WE	975682	**
94310	WE	94479	OM	95763	BS	975685	**
94311	WE	94481	SD	96100	TM	975686	**
94313	WE	94482	ML	96101	SN	975687	**
94316	SM	94488	CD	96110	CS	975688	**
94317	OM	94490	ML	96132	CS	975976	KT
94318	SD	94492	WE	96135	CS	975977	KT
94320	**	94495	TY	96164	CS	975991	CD
94322	ML	94497	ML	96165	CS	977077	**
94323	TY	94498	ML	96170	CS	977085	BH
94326	TY	94499	CD	96175	CS	977095	CS
94331	SD	94501	OM	96177	CP	977111	**
94332	TY	94504	TY	96178	CS	977112	**
94333	TY	94512	TY	96181	KT	977193	BH
94334	CD	94514	ML	96182	CS	977359	ZN
94335	TY	94515	ML	96191	CS	977390	CD
94336	TY	94517	CD	96192	CS	977399	NL
94337	WE	94518	BS	96210	CS	977526	SJ
94338	WE	94519	ML	96218	CS	977595	CD
94340	CD	94520	TY	96371	NP	977618	BY
94343	ML	94521	CD	96372	NP	977787	TH
94344	SM	94522	TY	96373	NP	977789	LU
94400	SD	94525	TY	96374	NP	977790	LU
94401	ML	94526	TY	96375	NP	977791	LU
94406	ML	94527	TY	96452	BR	977793	LU
94408	TY	94528	ML	96453	BR	977794	LU
94410	WE	94529	CD	96602	LM	977855	ZA
94411	SD	94530	ML	96603	LM	977905	EH
94412	ML	94531	TY	96604	LM	977944	TO

| 977945 TO | 977947 TO | 999503 OM | DS 70220 ** |
| 977946 TO | | | |

Note: Vehicles shown at ML may actually be located at Mossend Yard.

** Other locations:

6900	Cambridge Station Yard
6901	Cambridge Station Yard
92193	Preston Carriage Sidings
93180	Derby South Dock Siding
94320	Norwich Station
99014	Horsham Yard
99015	Horsham Yard
889026	Gloucester Yard
975379	Leeman Road, EY York
975497	Sudbrook
975535	Carnforth Bottom End Sidings
975681	Portobello
975682	Portobello
975685	Portobello
975686	Portobello
975687	Portobello
975688	Portobello
977077	Ripple Lane Yard
977111	Ripple Lane Yard
977112	Ripple Lane Yard
DS 70220	Western Trading Estate Siding, North Acton

Former Gatwick Express stock awaiting disposal:

Advertising livery: As **GX** but with a deep blue instead of a white lower bodyside, advertising Continental Airlines.

8208	**AL**	GB	PB	72507 (3412)	72643 (6040)	
8312	**GX**	GB	SN	72622 (6004)	72711 (6109)	72623 (6118)
8315	**GX**	GB	SN	72636 (6071)	72714 (6092)	72645 (5942)
Spare	**GX**	NR	AS	72608 (6077)		
Spare	**AL**	NR	AS	72641 (6079)		

NOTES

10. CODES

10.1. LIVERY CODES

Coaching stock vehicles are in Inter-City (light grey/red stripe/white stripe/dark grey) livery unless otherwise indicated. The colour of the lower half of the bodyside is stated first.

1	"One" (metallic grey with a broad black bodyside stripe. Pink, yellow, grey, pale green and light blue stripes at the unit/vehicle ends).
AR	Anglia Railways (turquoise blue with a white stripe).
AV	Arriva Trains (turquoise blue with white doors).
B	BR blue.
BG	BR blue & grey lined out in white.
BP	Blue Pullman (Nanking blue & white).
CC	BR Carmine & Cream.
CD	Cotswold Rail (silver with blue & red logo).
CH	BR Western Region/GWR chocolate & cream lined out in gold.
DR	Direct Rail Services (dark blue with light blue or dark grey roof).
DS	Revised Direct Rail Services (dark blue, light blue & green. "Compass" logo).
E	English Welsh & Scottish Railway (maroon bodyside & roof with gold band).
FB	First Group dark blue.
FD	First Great Western "Dynamic Lines" (dark blue with thin multi-coloured lines on lower bodyside).
FG	First Group Inter-City (indigo blue with gold, pink & white stripes).
FP	Old First Great Western (green & ivory with thin green & broad gold stripes).
FS	First Group (indigo blue with pink & white stripes).
FT	First Group "Dynamic Lines" {Pilot First Great Western} (varying blue with thin multi-coloured lines on lower bodyside).
G	BR Southern Region or SR green.
GC	Grand Central (black).
GN	Great North Eastern Railway (dark blue with a red stripe).
GX	Gatwick Express Inter-City (dark grey/white/burgundy/white).
HB	HSBC Rail (Oxford blue & white)
LN	LNER Tourist (green & cream).
M	BR maroon (maroon lined out in straw & black).
MA	Maintrain (blue).
MN	Midland Mainline (thin tangerine stripe on the lower bodyside, ocean blue, grey & white).
NR	Network Rail (blue with a red stripe).
O	Non-standard livery (see class heading for details).
P	Porterbrook Leasing Company (purple & grey or white).
PC	Pullman Car Company (umber & cream with gold lettering) lined out in gold.
RK	Railtrack (green & blue or plain blue).
RP	Royal Train (claret, lined out in red & black).

RR	Regional Railways (dark blue/grey with light blue & white stripes).
RV	Riviera Trains (Oxford blue & cream, lined out in gold).
SO	Serco Railtest (red & grey).
V	Virgin Trains (red with black doors extending into bodysides, three white lower bodyside stripes).
VN	Venice Simplon Orient Express "Northern Belle" (crimson lake & cream).
WX	Heart of Wessex Line promotional livery (cerise pink).
Y	Network Rail yellow.

10.2. OWNER CODES

24	6024 Preservation Society
62	The Princess Royal Locomotive Trust
92	City of Wells Supporters Association
A	Angel Trains
AW	Arriva Trains Wales
B1	Thompson B1 Locomotive Society
BB	Balfour Beatty Rail Plant
BE	Bert Hitchins
BK	The Scottish Railway Preservation Society
BS	Bressingham Steam Museum
CD	Cotswold Rail Engineering
CG	Cargo-D
CT	Central Trains
DE	Delta Rail (formerly AEA Technology Rail)
DG	Duke of Gloucester Steam Locomotive Trust
DM	Dartmoor Railway
DR	Direct Rail Services
E	English Welsh & Scottish Railway
EU	Eurostar (UK)
FG	First Group
FM	FM Rail (in administration)
GB	First GBRf
GS	The Great Scottish & Western Railway Company
GW	The Great Western Society
H	HSBC Rail (UK)
HI	Hitachi
IR	Ian Riley Engineering
J	Fastline (Jarvis Rail)
JC	John Cameron
JH	Jeremy Hosking
LW	London & North Western Railway Company
MH	Mid Hants Railway
MM	Midland Mainline
MN	Merchant Navy Locomotive Preservation Society
NE	North Eastern Locomotive Preservation Group
NM	National Railway Museum
NR	Network Rail
NY	North Yorkshire Moors Railway
P	Porterbrook Leasing Company
RA	Railfilms
RP	Rampart Carriage & Wagon Works
RV	Riviera Trains
SH	Scottish Highland Railway Company
SM	Siemens Transportation
SO	Serco Railtest
ST	Sovereign Trains (sister company to Grand Central)
SV	Severn Valley Railway
VS	Venice-Simplon Orient Express
VT	Vintage Trains
WC	West Coast Railway Company
WS	Wrexham Shropshire & Marylebone Railway
WT	Wessex Trains

10.3. OPERATOR CODES

The two letter operator codes give the current operator. This is the organisation which facilitates the use of the coach and may not be the actual train operating company which runs the train on which the particular coach is used. If no operator code is shown then the vehicle is not at present in use.

62	The Princess Royal Locomotive Trust
1	"One"
AW	Arriva Trains Wales
BB	Balfour Beatty Rail Plant
BK	The Scottish Railway Preservation Society
CD	Cotswold Rail
CG	Cargo-D
CT	Central Trains
DR	Direct Rail Services
E	English Welsh & Scottish Railway
EU	Eurostar (UK)
FL	Freightliner
GB	First GBRf
GC	Grand Central
GN	Great North Eastern Railway
GS	The Great Scottish & Western Railway Company
GW	First Great Western
ME	Merseyrail Electrics
MH	Mid Hants Railway
MM	Midland Mainline
MR	ECT Mainline Rail
NY	North Yorkshire Moors Railway
RP	Royal Train
RV	Riviera Trains
SH	Scottish Highland Railway Company
SO	Serco Railtest
SR	First ScotRail
VI	Victa Westlink Rail
VS	Venice-Simplon Orient Express
VT	Vintage Trains
VW	Virgin West Coast
WC	West Coast Railway Company
WT	Wessex Trains

10.4. ALLOCATION & LOCATION CODES

Code	Depot	Operator
AC	Aberdeen Clayhills	*Storage location only*
AP*	Ashford Rail Plant	Balfour Beatty Rail Plant
AS*	Alley's, Studley (Warwickshire)	*Storage location only*
BD	Birkenhead North	Merseyrail Electrics
BH	Barrow Hill (Chesterfield)	Barrow Hill Engine Shed Society
BI	Brighton Lovers Walk	Southern
BK	Bristol Barton Hill	EWS
BN	Bounds Green (London)	GNER
BQ	Bury (Greater Manchester)	East Lancashire Railway/Ian Riley
BR*	MoD DSDC Bicester	Ministry of Defence
BS	Bescot (Walsall)	EWS
BT	Bo'ness (West Lothian)	Bo'ness & Kinneil Railway
BY	Bletchley	Silverlink
CD	Crewe Diesel	EWS
CE	Crewe International Electric	EWS
CF	Cardiff Canton	Arriva Trains Wales/Pullman Rail
CJ	Clapham Yard (London)	South West Trains
CO	Cranmore (Somerset)	East Somerset Railway
CP	Crewe Carriage	London & North Western Railway Co.
CR	Crewe Gresty Lane	Direct Rail Services
CS	Carnforth	West Coast Railway Company
CT*	MoD Caerwent AFD (Caldicot)	Ministry of Defence
CV	Coalville Mantle Lane	*Storage location only*
DI	Didcot Railway Centre	Great Western Society
DR	Doncaster	EWS
DW*	Doncaster West Yard	*Storage location only*
DY	Derby Etches Park	Midland Mainline
EC	Edinburgh Craigentinny	GNER
EH	Eastleigh	EWS
EM	East Ham (London)	c2c Rail
GL	Gloucester Horton Road	Cotswold Rail
GW	Glasgow Shields Road	First ScotRail
HE	Hornsey (London)	First Capital Connect
HM	Healey Mills (Wakefield)	EWS
HT	Heaton (Newcastle)	Northern
IL	Ilford (London)	"One"
IS	Inverness	First ScotRail
KM	Carlisle Kingmoor	Direct Rail Services
KR	Kidderminster	Severn Valley Railway
KT	MoD Kineton (Warwickshire)	Ministry of Defence
LA	Laira (Plymouth)	First Great Western
LM	Long Marston (Warwickshire)	St. Modwen Properties
LU*	MoD Ludgershall	Ministry of Defence
MA	Manchester Longsight	West Coast Traincare
MG	Margam (Port Talbot)	EWS
ML	Motherwell (Glasgow)	EWS
MQ*	Meldon Quarry (Okehampton)	Dartmoor Railway

NC	Norwich Crown Point	"One"
NL	Neville Hill (Leeds)	Midland Mainline/Northern
NP	North Pole International (London)	Eurostar (UK)
NY	Grosmont (North Yorkshire)	North Yorkshire Moors Railway
OM	Old Oak Common Carriage (London)	Riviera Trains/EWS
OO	Old Oak Common HST	First Great Western
OY	Oxley (Wolverhampton)	West Coast Traincare
PB	Peterborough Yards	EWS/First GBRf
PM	St. Philip's Marsh (Bristol)	First Great Western
PZ	Penzance Long Rock	First Great Western
RL	Ropley (Hampshire)	Mid-Hants Railway
SD*	Stoke Gifford Yard (Bristol Parkway)	*Storage location only*
SI	Soho (Birmingham)	Maintrain/Central Trains
SJ*	Severn Tunnel Junction	EWS
SK	Swanwick Junction (Derbyshire)	Midland Railway-Butterley
SL	Stewarts Lane (London)	Gatwick Express/VSOE
SM*	Swansea Maliphant Sidings	*Storage location only*
SN*	MoD DERA Shoeburyness	Ministry of Defence
SO*	Southall (Greater London)	Jeremy Hosking
TE	Thornaby (Middlesbrough)	EWS
TH*	Pershore Airfield, Throckmorton, Worcs.	*Storage location only*
TJ	Tavistock Junction Yard (Plymouth)	Colas Rail
TM	Tyseley Locomotive Works	Birmingham Railway Museum
TN	Thornton (Fife)	John Cameron
TO	Toton (Nottinghamshire)	EWS
TW*	Tonbridge West Yard	First GBRf
TY	Tyne Yard (Newcastle)	EWS
WB	Wembley (London)	Alstom
WD	Wimbledon (London)	South West Trains
WE	Willesden Brent Sidings	*Storage location only*
YK	National Railway Museum (York)	Science Museum
ZA	RTC Business Park (Derby)	Serco Railtest/AEA Technology
ZB	Doncaster Works	Wabtec
ZC	Crewe Works	Bombardier Transportation
ZD	Derby, Litchurch Lane Works	Bombardier Transportation
ZG	Eastleigh Works	Knights Rail Services/Wabtec
ZH	Springburn Works (Glasgow)	Railcare
ZI	Ilford Works	Bombardier Transportation
ZJ	Marcroft, Stoke	Turners
ZK	Kilmarnock Works	Hunslet-Barclay
ZN	Wolverton Works	Railcare
ZR	York (former Thrall Works)	Network Rail

*= unofficial code.

ABBREVIATIONS

AFD	Air Force Department
DERA	Defence Evaluation & Research Agency
DSDC	Defence Storage & Distribution Centre